Systematics, Breeding and Seed Production of Potatoes

RUSSIAN TRANSLATIONS SERIES

1. K.Ya. Kondrat'ev et al. (editors): *USSR/USA Bering Sea Experiment*
2. D.V. Nalivkin: *Hurricanes, Storms and Tornadoes*
3. V.M. Novikov (editor): *Handbook of Fishery Technology, Volume 1*
4. F.G. Martyshev: *Pond Fisheries*
5. R.N. Burukovskii: *Key to Shrimps and Lobsters*
6. V.M. Novikov (editor): *Handbook of Fishery Technology, Volume 4*
7. V.P. Bykov (editor): *Marine Fishes*
8. N.N. Tsvelev: *Grasses of the Soviet Union*
9. L.V. Metlitskii et al.: *Controlled Atmosphere Storage of Fruits*
10. M.A. Glazovskaya: *Soils of the World* (2 volumes)
11. V.G. Kort & V.S. Samoilenko: *Atlantic Hydrophysical Polygon-70*
12. M.A. Mardzhanishvili: *Seismic Design of Frame-Panel Buildings and Their Structural Members*
13. E'.A. Sokolenko (editor): *Water and Salt Regimes of Soils: Modeling and Management*
14. A.P. Bocharov: *A Description of Devices Used in the Study of Wind Erosion of Soils*
15. E.S. Artsybashev: *Forest Fires and Their Control*
16. R.Kh. Makasheva: *The Pea*
17. N.G. Kondrashova: *Shipboard Refrigeration and Fish Processing Equipment*
18. S.M. Uspenskii: *Life in High Latitudes*
19. A.V. Rozova: *Biostratigraphic Zoning and Trilobites of the Upper Cambrian and Lower Ordovician of the Northwestern Siberian Platform*
20. N.I. Barkov: *Ice Shelves of the Antarctic*
21. V.P. Averkiev: *Shipboard Fish Scouting and Electronavigational Equipment*
22. D.F. Petrov (Editor-in-Chief): *Apomixis and Its Role in Evolution and Breeding*
23. G.A. Mchedlidze: *General Features of the Paleobiological Evolution of Cetacea*
24. M.G. Ravich et al.: *Geological Structure of Mac. Robertson Land (East Antarctica)*
25. L.A. Timokhov (editor): *Dynamics of Ice Cover*
26. K.Ya. Kondrat'ev: *Changes in Global Climate*
27. P.S. Nartov: *Disk Soil-Working Implements*
28. V.L. Kontrimavichus (Editor-in-Chief): *Beringia in the Cenozoic Era*
29. S.V. Nerpin & A.F. Chudnovskii: *Heat and Mass Transfer in the Plant–Soil–Air System*
30. T.V. Alekseeva et al.: *Highway Machines*
31. N.I. Keenin et al.: *Agricultural Machines*
32. V.K. Rudnev: *Digging of soils by Earthmovers with Powered Parts*
33. A.N. Zelenin et al.: *Machines for Moving the Earth*
34. *Systematics, Breeding and Seed Production of Potatoes*
35. D.S. Orlov: *Humus Acids of Soils*
36. M.M. Severnev (editor): *Wear of Agricultural Machine Parts*
37. Kh. A. Khachatryan: *Operation of Soil-Working Implements in Hilly Regions*
38. L.V. Gyachev: *Theory of Surfaces of Plow Bottoms*
39. S.V. Kardashevskii et al.: *Testing of Agricultural Technological Processes—A Systems Approach*

BULLETIN OF APPLIED BOTANY, GENETICS AND BREEDING
VOL. 62, NO. 1

Systematics, Breeding and Seed Production of Potatoes

RUSSIAN TRANSLATIONS SERIES
34

1986
A.A. BALKEMA/ROTTERDAM

Translation of:
Sistematika, Selektsiya i Semenovodstvo Kartofelya (Trudy po Prikladnoi Botanike, Genetike i Selektsii, Tom 62, Vypusk 1).
Leningrad, 1978

Translator: Dr. A.K. Dhote
General Editor: Dr. V.S. Kothekar

ISBN 90 6191 452 3

Printed in India at Printsman Press, Faridabad

Contents

TRUDY PO PRIKLADNOI BOTANIKE, GENETIKE I SELEKTSII

1978 Vol. 62, No. 1

Systematics of the Potato

S.M. Bukasov
Academician VASKhNIL

This paper reviews the systematics of the potato species: from Dunal to Bitter to recent authors. Potato species are included in the section Tuberarium (Dun.) Buk. This section is divided into subsections and series. The concept of potato species is presented according to N.I. Vavilov and S.V. Yuzepchuk's interpretation. Morphological, cytological, geographic, and phylogenetic principles are used to establish different taxa.

The taxonomic significance of characters—the hierarchy—is estimated according to N.I. Vavilov's concept. Hawkes and Dodds' errors in the systematics of cultivated potato species are discussed.

Hawkes contravened the principles underlying the systematics of the cultivated potato species. He disregarded the principles of geography and the geographic range of the species. He considered, without justification, the dormancy period of the tubers as well as the number of chromosomes to be of primary importance, regardless of the phylogeny of the species. Dodds eliminated the division of the cultivated potato into species, considering them to be only cultivars of the lone species Solanum tuberosum. The problems of the origin of the cultivated potato from the wild ones are also discussed.

The present article is based on the concept of N.I. Vavilov on the systematics of the species of Linnaeus. Only the interpretation of the volume of the taxon is enlarged and relates to the species in a narrow sense, as we know Linnaeus's *Solanum tuberosum* L.; and a series of species and subsections has been established. This was done because genus *Solanum* L. consists of an excessively large number of species united together on the basis of insignificant characters which resulted in the exclusion of *Lycopersicon* and *Melongena* from the genus *Solanum* and their establishment as independent genera. The eminent modern solanoloist G. Bitter [14] subdivided genus *Solanum* into sections and one of these, viz. *Tuberarium* (Dun.) Bitt., included tuber-forming species, i.e. the potato.

On the basis of the concept of N.I. Vavilov relating to the importance of the geographical principle and the distinction of separate geographic ranges in the systematics of a species, we have subdivided section *Tuberarium* into 25 subsections following the principle of morphological separation by dissection of the leaf and the type of glandules.

THE SYSTEMATICS OF SUBSECTIONS AND SERIES OF WILD AND CULTIVATED SPECIES OF THE POTATO OF SECTION TUBERARIUM (DUN.) BUK., GENUS SOLANUM L.

In a manner similar to the series of species, the subsections are arranged according to the principle of their phylogeny and geography. Species are arranged alphabetically within the limits of each series.

SOUTH AMERICAN SUBSECTIONS

1. **Berthaultianum** Buk. Chromosomes 24.

It is distinguished from other subsections by the presence of unicellular glandules on multi-cellular stalks; beside these, there are glandules of two more types on unicellular stalks, viz., tetra- and hexacellular glandules.

Geographic range: Bolivia, 2000–2700 meters above mean sea level.

Series *Berthaultiana* Buk.

S. Berthaultii Hawk.
S. zudanense Card.

2. **Andinum** Buk. Chromosomes 24, 48.

Corolla pentangular; berries deciduous.

Geographic range: Upper belt of the Andes, 2000–5000 meters above m.s.l.

Series *Transaequatorialia* Buk.

S. aracc-papa Juz.
S. brevicaule Bitt.
S. bukasovii Juz.
S. calcense Bitt.
S. famatinae Bitt. et Wittm.
S. gourlayi Hawk.
S. kurtzianum Bitt. et Wittm.
S. canasense Hawk.
S. candolleanum Berth.
S. catarthrum Juz.
S. coelestipetalum Vargas.
S. oplocense Hawk.
S. ruiz-cevallosii Card.
S. setulosistylum Bitt.

S. lechnoviczii Hawk.
S. leptophyes Bitt.
S. liriunianum Card. et Hawk.
S. macolae Buk.
S. multidissectum Hawk.
S. multiinterruptum Bitt.
S. neohawkesii Ochoa
S. soukupii Hawk.
S. sparsipilum (Bitt.) Juz. et Buk.
S. spegazzinii Bitt.
S. sucrense Hawk., 2n=48
S. velascanum Bitt.
S. vernei Bitt. et Wittm.

Series *Cisaequatorialia* Buk., 2n=24.

Geographic range: north of the Equator in Colombia, 2400–2900 meters above m.s.l.

S. andreanum Baker.

Series *Piurana* Hawk.

Geographic range: from Peru to southern Colombia, 2100–4100 meters above m.s.l.

S. chiquidenum Ochoa
S. chomatophilum Bitt.
S. cuzcoense Ochoa
S. marinasense Vargas
S. pampasense Hawk.
S. piurae Bitt.

Series *Cuneoalata* Hawk., 2n=24.

Geographic range: Chile, Bolivia and Argentina, 2500–3800 meters above m.s.l.

S. infundibuliforme Phil.
S. xerophilum Hawk. et Card.

Series *Minutifoliola* Corr., 2n=24.

Geographic range: Peru and Ecuador, 1500–2900 meters above m.s.l.

S. cajamarcense Ochoa
S. Minutifoliolum Ochoa

Series *Megistacroloba* Card. et Hawk., 2n=24.

Geographic range: Argentina, Bolivia and Peru, 2500–3900 meters above m.s.l.

S. megistacrolobum Bitt.
S. raphanifolium Card. et Hawk.
S. sanctae-rosae Hawk.

Series *Lignicaulia* Buk., 2n=24.

Geographic range: Peru, 2800–3800 meters above m.s.l.

S. lignicaule Vargas.

Series *Ingaefolia* Ochoa, 2n=48.

Geographic range: Northern Peru, Ecuador, Southern Colombia, 2800–3300 meters above m.s.l.

S. ingaefolium Ochoa
S. tuguerrense Hawk.

3. **Gigantophyllum** Buk. Chromosomes 24.
Leaves large, simpie.
Geographic range: Bolivia and Argentina, 1100–3400 meters above m.s.l.

Series *Simpliciora* Buk.

S. gandarillasii Card.
S. gigantophyllum Bitt.
S. microdontum Bitt.
S. simplicifolium Bitt.
S. venturii Hawk. et Hjert.
S. virgultorum (Bitt.) Card. et Hawk.

4. **Acaule** Buk. Chromosomes 24, 48, 72.
Plants rosette-shaped. Inflorescence with one whorl on the vestige of peduncle, berries non-deciduous.
Geographic range: from Argentina to Peru, 3500–5000 meters above m.s.l.

Series *Alticola* Buk., 2n=24.

Geographic range: Bolivia and Peru, 3000–4100 meters above m.s.l.

S. alticolum Bitt.
S. boliviense Dun.
S. ellipsifolium Card. et Hawk.
S. sogarandinum Ochoa
S. tilcarense Hawk.
S. toralapanum Card. et Hawk.

Series *Acaulia* Juz., 2n=48, 72.

Geographic range: Argentina, Bolivia and Peru, 3500–5000 meters above m.s.l.

S. acaule Bitt.
S. aemulans Bitt. et Wittm.
S. albicans (Ochoa) Buk., 2n=72
S. depexum Juz.
S. punae Juz.
S. Schreiteri Buk.

5. **Conicibaccatum** Buk. Chromosomes 24, 48, 72.
Berries conical, deciduous.
Geographic range: from Bolivia to Venezuela, 1200–4500 meters above m.s.l.

Series *Conicibaccata* Bitt.

S. ayacuchense Ochoa, 2n=24.
S. colombianum Dun., 2n=48.
S. huancabambense Ochoa, 2n=24.
S. moscopanum Hawk, 2n=72.
S. otites Dun., 2n=48.
S. santolallae Vargas, 2n=24.
S. urubambae Juz., 2n=24.
S. venezuelicum Buk., 2n=24.
S. violaceimarmoratum Bitt., 2n=24.

6. **Capsicibaccatum** Buk. Chromosomes 24.
Corolla stellate. Berries capsicum shaped.
Geographic range: Bolivia, 2000–4000 meters above m.s.l.

Series *Circaeifolia* Hawk.

S. capsicibaccatum Card.
S. circaeifolium Bitt.

7. **Tarijense** Buk. Chromosomes 24.
Pubescence thick, glandular, Corolla stellate. Berries round.
Geographic range: Bolivia and Argentina, 2000–3000 meters above m.s.l.

Series *Tarijensa* Corr.

S. tarijense Hawk.

8. **Orientale** Buk. Chromosomes 24.
Corolla stellate, white. Berries round, deciduous.
Geographic range: Eastern slopes of the Andes; Bolivia and Argentina; Paraguay, Uruguay, Brazil, below 3200 up to 500 meters above m.s.l.

Series *Jungasensa* Corr.

S. trigalense Card.
S. yungasense Hawk.

Series *Glabrescentia* Buk.

S. boergeri Buk.
S. chacoense Bitt.
S. cuevoanum Card.
S. dolichostigma Buk.
S. garciae Juz. et Buk.
S. gibberulosum Juz. et Buk.
S. horovitzii Buk.
S. emmeae Juz. et Buk.
S. knappei Juz. et Buk.
S. schickii Juz. et Buk.
S. subtilius Bitt.

9. **Commersonianum** Buk. Chromosomes 24, 36.
Plants rosette-shaped. Berries cordate.
Geographic range: La Plata lowland of Argentina and Uruguay, below 500 meters above m.s.l.

Series *Commersoniana* Buk.

S. commersonii Dun., 2n = 24, 36.
S. henryi Buk., 2n = 24.
S. mechonguense Buk., 2n = 24.
S. millanii Buk., 2n = 36.
S. ohrondii Corr., 2n = 24.
S. pyriforme Buk., 2n = 24.
S. rionegrinum Buk., 2n = 24.

10. **Vavilovianum** Buk. Chromosomes 24.
The leaves smell like *Lycopersicon*.
Geographic range: Lima, Peru, below 100 meters above m.s.l.

Series *Vaviloviana* Buk.

S. vavilovii Juz.
S. wittmackii Bitt.

11. **Pacificum** Buk. Chromosomes 24, 36, 48.

Series *Collina* Buk., 2n = 24, 36.

Corolla white or bluish violet.
Geographic range: coastal areas of Peru and north Chile.

S. chancayense Ochoa, 2n=24.
S. immite Dun., 2n=24.
S. maglia Schlechtd., 2n=24, 36.
S. medians Bitt., 2n=24.
S. mochiquense Ochoa, 2n=24.
S. neoweberbaueri Wittm, 2n=24.
S. weberbaueri Bitt., 2n=24.

Series *Tuberosa* (Rydb.) Buk., 2n=48.

Geographic range: coastal regions of central Chile along with adjoining islands.

S. guaytecarum (Bitt.) Buk.
S. leptostigma Juz.
S. molinae Juz.
S. zykinii Lechn.

NORTH AMERICAN SUBSECTIONS

12. **Oxycarpum** Buk. Chromosomes 48.
Berries conical, deciduous.
Geographic range: from Panama and Costa Rica to Mexico, 1500–3000 meters above m.s.l.

Series *Oxycarpa* Buk.

S. agrimonifolium Rydb.
S. oxycarpum Schiede.

13. **Verrucosum** Buk. Chromosomes 24.
Plants dwarf. Corolla pentangular, bluish-purple.
Berries ridged, non-deciduous or persistent.
Geographic range: central Mexico, 2400–3200 meters above m.s.l.

Series *Verrucosa* Buk.

S. verrucosum Schlechtd.

14. **Arcticum** Buk. Chromosomes 60, 72.
Plants squat. Corolla deciangular. Berries persistent oval and long.
Geographic range: central and southern Mexico, Guatemala, 2650–3800 meters above m.s.l.

Series *Demissa* Buk.

S. brachycarpum Corr., 2n=72
S. demissum Lindl., 2n=72
S. edinense A. Sutton, 2n=60
S. hougasii Corr., 2n=72.
S. guerreroense Corr., 2n=72.
S. iopetalum (Bitt.) Hawk., 2n=72.
S. salamanii Hawk., 2n=60.
S. semidemissum Juz., 2n=60.

15. **Longipedicellatum** Buk. Chromosomes 36, 48.
Peduncles long. Berries deciduous.
Geographic range: from southern Mexico to adjoining states of the USA, 1500–3600 meters above m.s.l., rarely 3800 meters above m.s.l.

Series *Longipedicellata* Buk.

S. ajuscoense Buk.
S. antipoviczii Buk.
S. fendleri A. Gray
S. hjertingii Hawk.
S. malinchense Hawk.
S. papita Rydb.
S. polytrichon Rydb.
S. stoloniferum Schlechtd.
S. tlaxcalense Hawk.
S. vallis mexici Juz., 2n = 36.

16. **Boreale** Buk. Chromosomes 48.

Corolla pentangular, purple. Berries deciduous.

Geographic range: Mexico to the state of Texas in the USA, 2000–3300 meters above m.s.l.

Series *Borealia* Corr.

S. leptosepalum Corr.
S. macropilosum Corr.
S. wightianum Rydb.

17. **Polyadenium** Buk. Chromosomes 24.

Numerous glandules with repulsive smell.

Geographic range: central Mexico, 1900–2900 meters above m.s.l.

Series *Polyadenia* Buk.

S. lesteri Hawk.
S. polyadenium Greenm.

18. **Pinnatisectum** Buk. Chromosomes 24.

Leaves without intermediate lobes. Corolla stellate, white. Berries round, deciduous.

Geographic range: from central Mexico to adjoining states of the USA, 1500–2400 meters above m.s.l.

Series *Pinnatisecta* Rydb.

S. brachistotrichum (Bitt.) Rydb.
S. jamesii Torr.
S. pinnatisectum Dun.
S. sambucinum Rydb.
S. stenophyllidium Bitt.

19. **Trifidum** Buk. Chromosomes 24.

Leaves trisected, i.e. with one pair of basic lobes. Corolla stellate, white.

Geographic range: central and western Mexico, 2200–2320 meters above m.s.l.

Series *Trifida* Corr.

S. michoacanum (Bitt.) Rydb.
S. trifidum Corr.

20. **Cardiophyllum** Buk. Chromosomes 24, 36.

Leaf lobes cordate. Corolla stellate.

Geographic range: central and northern Mexico, 1500–2600 meters above m.s.l.

Series *Cardiophylla* Buk.

S. cardiophyllum Lindl., 2n = 24.
S. ehrenbergii (Bitt.) Rydb., $2n = 24$.
S. coyoacanum Buk., 2n = 36.

21. **Bulbocastanum** Buk. Chromosomes 24, 36.
Leaves entire, small. Corolla stellate, with narrow petals.
Geographic range: from central Mexico to Guatemala, 1500–3800 meters above m.s.l.

Series *Bulbocastana* Rydb., 2n = 24, 36.
Geographic range: Mexico, 1500–2100 meters above m.s.l.

S. bulbocastanum Dun.

Series *Clara* Graham et Dionne, 2n = 24.

Geographic range: Guatemala and Mexico, 3200–3800 meters above m.s.l.

S. clarum Corr.

22. **Morelliforme** Buk. Chromosomes 24.
Plants epiphytic. Leaves entire. Corolla stellate, small.
Geographic range: from east-central Mexico to Guatemala, 1600–2700 meters above m.s.l.

Series *Morelliformia* Hawk.

S. morelliforme Bitt. et Muench.

CULTIVATED NATIVE SPECIES OF THE SOUTH AMERICAN POTATO

23. **Andigenum** Buk. Chromosomes 24, 36, 48.

Series *Andigena*.

Diploids

S. ajanhuiri Juz. et Buk.
S. boyacense Juz. et Buk.
S. canarense Buk.
S. goniocalyx Juz. et Buk.
S. kesselbrenneri Juz. et Buk.
S. multijugum Buk. et Bavyko.
S. phureja Juz. et Buk.
S. rybinii Juz. et Buk.
S. stenotomum Juz. et Buk.

Triploids

S. chaucha Juz. et Buk.
(= *S. surimana*)
S. cuencanum Buk.
S. mammilliferum Juz. et Buk.
S. tenuifilamentum Juz. et Buk.

Tetraploid

S. andigenum Juz. et Buk.

24. **Subacaule** Buk. Chromosomes 36, 40.

Series *Subacaulia* Buk.

S. juzepczukii Buk., 2n = 36.
S. curtilobum Juz. et Buk., 2n = 60.

25. **Chilotanum** Buk. Chromosomes 48.

Series *Chilotana* Buk. Chromosomes 48.

S. chilotanum Hawk.

The history of the systematics of the potato began with the establishment of the taxon *Tuberarium* in 1852 by M. Dunall [22] and C. Walpers [42] and it has significant controversial examples due to the violation of some basic taxonomic principles.

In the beginning the taxon *Tuberarium* was described by Dunall (1852) as a subsection including tuberous as well as non-tuberous species of *Solanum* L. S. Dannert [20] considers that according to Walpers the section including tuberous species should be termed "potato". Bitter [14] committed a mistake by including in the section *Tuberarium* (Dun.) Bitt. subsection *Basarthrum* comprising non-tuberous species and subsection *Hyperbasarthrum* Bitt. comprising tuberous species as well as a number of species not developing tubers. Part of the latter species was listed by Ridberg in the series *Juglandifolia* while the remaining ones were included in the series *Etuberosa* by S.V. Yuzepchuk. All non-tuberous species were excluded by us [3] from the section *Tuberarium* (Dun.) Buk.

Almost 250 years have passed since the first description of the cultivated species of the potato *S. tuberosum* L. and its inclusion among species of genus *Solanum*. J. Molina [29] mentions the origin of wild species of the potato. To date about two hundred wild species of the potato have been described and during this period these species have been collected by hundreds of botanists for their herbaria and thousands of leaves of these species have been preserved in many countries. Major collections of herbarial material were made during the last few decades after the first Soviet expeditions revealed the diversity of potato species and negated the earlier concepts of the poor specific diversity of the potato.

After the pioneer Soviet expeditions to South and Central America in 1925–1930, a number of foreign expeditions were directed to search for new species and new places of origin of the wild potato. Beside leaves, the botanists also collected tubers and seeds which provided an opportunity to study these species in a live state. However, collection of live plant material is possible only during the limited period of fruit-bearing or formation of tubers. Moreover, preservation and transportation of tubers is more complicated than herbarial material. Therefore, in the known collections of wild species of the potato we find only a few specimens, numbering not more than two hundred, against thousands of herbarial specimens.

Many wild species of the potato are known only on the basis of herbarial material and they are not collected alive even by the local botanists of Latin America. Many species have perished or are under threat of complete extinction due to cultivation of land, for example species from Chile such as *S. molinae*, *S. leptostigma*, *S. guaytecarum* (Bitt.) Buk., as well as other species which were collected only once but were never found again by subsequent botanists.

In the first monograph on the potato species systematized by Dunall in 1852, less than twenty species are mentioned—those which were known at that time. The systematics and description became more complicated with the discovery of new species because new taxonomic characters had been introduced which were used for description (without metrological evaluation) by a few and were restricted mostly to the characters of the leaf and corolla.

The first detailed description of the species was presented by Bitter [14]. In recent years D. Correll [19] has given a detailed description of all species of the potato. Detailed descriptions of some species discovered by S.V. Yuzepchuk and J. Hawkes [24] add much to the list of taxonomic characters of potatoes. These authors dealt with large taxonomic units such as groups and series. According to S.V. Yuzepchuk [12], there are no fundamental differences between the wild and cultivated species from the taxonomic point of view. In order to design the phylogenetic system of a cultivated plant we must be guided by the following principles: to establish the close relatives of the cultivated plant, and its geographic range and distribution at present and in the past; to establish where and how the potato crop emerged, and who were the people that developed this crop; in which direction did the initial plants which were taken for cultivation change; were the initial forms crossed with each other; what was the role played by natural and artificial selection and which parts of the plant were affected by such selection.

We visualize the ideal, rational system of the cultivated plant in this form: from one side it must be based on the system of wild species and should be as if developed from their background, and from other side it should never be linked with the latter because of the non-feasibility of comparing the measurements of wild and cultivated species, but it is possible to separate the latter ones more distinctly.

At one time we were required to complete the work on differentiation of *Solanum tuberosum* s. l. into species [10]. Initially 12 of them were distinguished, but subsequent work added many more species to them and their number increased to 20. One of the established species, viz. *S. andigenum* Juz. et Buk., is quite polymorphic.

The common origin of some 24-chromosome potatoes, such as *S. rybinii*, *S. canarense*, *S. goniocalyx*, *S. stenotomum* and *S. phureja* occupying a comparatively small geographic range, is not yet proved. It is not proved for 48-chromosome potatoes (*S. andigenum* and *S. tuberosum*); in fact, on the contrary, there are grounds for presuming that they originated from absolutely different wild forms. Some 36-chromosome potatoes, such as *S. tenuifilamentum*, *S. mammilliferum*, *S. chaucha* and *S. juzepczukii*, are distinctly hybridogenic and of absolutely different origin. Obviously, they should neither be combined with each other, nor be affiliated to some non-hybrid forms [12].

Acknowledging the usefulness of descriptions of species by many authors, we cannot help pointing out their shortcomings particularly as pertaining to the analysis of the taxonomic importance of morphological characters. Therefore, the present work has a task, viz. to conduct an analysis of the taxonomic characters of potato species on the basis of many years' study of the world collection of VIR grown near Leningrad.

VARIABILITY OF CHARACTERS

Herbaria provide an opportunity to establish the great variability of characters in specimens from different places. These specimens became available during the past few years. Where the species was described from only one known place of occurrence, its separation as an independent species becomes valid if the characters absent in species of the same geographic range or from adjoining geographic ranges are present. Therefore, the need emerges to establish, according to N.I. Vavilov, a "hierarchy of characters" which is the basis for separation of large taxa such as series and subsections.

Ecological variability of a character was established during a study of wild species of potatoes from Bolivia and Peru which stimulated us to study and analyze, gradually changing characters in species close to each other in respect of their geographic ranges. In the case of distant locations of the same species, the difference in ecological conditions of such geographic ranges created the wide variability of characters which provided a basis for describing plants from different localities as independent species. This happened many times while describing species represented by solitary specimens. Subsequent expeditions yielded connecting links which were fairly adequate in combining two or more species (which were earlier described as independent species) into one species. As an addition to the criteria of ecological variability of species, it is necessary to study these characters in progeny developed through self-pollination.

D. Ugent [37] studied wild species of potatoes in their natural habitat in Bolivia and Peru for their ecological as well as genetic variability under induced (artificial) self-pollination. Taking *S. brevicaule* as a typical species with three subspecies, viz. *brevicaule*, *canasense* and *bukasovii*, he united about 15 "minor" species into one unified species. However, the problem arises on the boundaries of the species and subspecies which remain controversial even today if we confine ourselves only to morphological and geographical differences.

The ability of species to be crossed and the fertility of the hybrid progeny are subjects of special investigation, similar to the study of genomes and cytogenetics which does not fall into the purview of the "pure" morphological taxonomy and is closer to phylogenetic problems.

The peculiarity of a character or its extreme peripheral position in wide variability, as well as the value of this variability, help us to establish the hierarchy of the taxonomic character. Examples of the peculiarity of characters are: repulsive smell (*S. polyadenium*) or unicellular glandules (*S. berthaultii*). Extreme variations regarding dissection of the leaf (from entire to highly dissected with intermediate lobes) are the characters of a series. Extreme variations in respect to the dissection of the corolla (from round to stellate) serve to distinguish the subsections. Small differences in absolute numerical expression can become large in relative expression.

The taxonomic value of the inflorescence, flower, chromosome number and physiological characters (disease resistance, frost resistance), shattering of berries as well as biochemical characters, can be established only on live plants.

Anatomical characters which are established through microscopic studies supplement the anatomical characters which are established visually (or through touch or smell) and which do not require magnification by microscope or preparation of special anatomical slides, such as while describing the anatomical structure of leaf, stem, starch grains, tubers, structure of the articulation of peduncle and pedicel.

The taxonomic value of a character of the species is determined by its constantness: 1) frequency of occurrence in its geographic range, 2) radius of the geographic range, 3) genetic nature (dominance) as revealed through sexual reproduction. The presence of all these factors determines the value of the character.

If the character has small variability, such as the highly laminated rosette of leaves in *S. depexum,* acquires a unique position as the specific taxonomic character. For example, *S. depexum* is characteristic of the entire series *Acaulia*. For other species with not so prominent a rosette, it is characteristic to observe the expression of this character in a few subsequent variations in the lower part of the stem also.

Similar instances refer to quantitative characters. But qualitative characters appear in some species as unique ones inherent only in a solitary species from *Tuberarium* and these characters are not repeated in any of the other species. Repulsive smell (in *S. polyadenium* and the entire series *Polyadenia*) and unicellular glandules (series *Berthaultiana*) are examples of such characters.

Initially, morphological characters were used in the history of plant systematics. This was inevitable when the botanists were constrained to conduct studies on herbarial materials only. Description of the herbarial specimens was incomplete and was confined to much bigger differences between species. Such were the descriptions by C. Linnaeus [27], and later on by Dunall and other authors of the last century. Variability of species during ontogenesis was not taken into consideration while describing the characters of the species. While using (studying) live plants for systematics, their biological characters also proved to be of great importance and they are used in the present work, also.

In the classic descriptions of species, for example those by S.V. Yuzepchuk, we find a detailed description of differences in the pubescence of all plant organs, viz., stem, pedicel, rachis and veins of the leaf (main, their branches, mesophyll), as well as the dorsal and ventral surfaces of the leaf, flower, peduncle and calyx. Pubescence is duly mentioned while describing the corresponding organ of the plant. Moreover, organs whose pubescence is the specific distinguishing character of the species are not identified.

Therefore, we must describe the pubescence of all organs in one paragraph, singling out the organ in which pubescence is inherent according to the peculiarities (shape, location, length and density) of the species being described. Moreover, it is also necessary to describe the differences in the pubescence of the lower and upper parts of the peduncle and branches of inflorescence as well as its ontogenesis. While describing the non-shattering nature of the berries, it is necessary to establish the constantness of this character at the time of complete maturity of the berries. The cellular layer dividing the articulation of the peduncle develops differently in different species. Therefore, immature berries may remain persistent even after drying in the bunches of berries taken from a plant. For example, it is still not clear whether in fact mature berries of *S. vernei* and *Alticola* do fall. It is easy to describe berries of *Tuberosa*, *Andigena*, *Longipedicellata Glabrescentia*. At the same time there are differences in the periods of falling off of the berries between species of the same series as well as different forms of one species and between varieties of the potato.

The mathematical method in the systematics of the potato. The systematics of a cultivated plant must be linked with the systematics of its wild relatives, and hence it is based on general principles of botanical systematics, but taking into consideration the specificity of the cultivated plant and the systematics of populations. At the initial stages of systematics a purely descriptive morphological method prevailed, without the inclusion of mathematical indices for the characters being described. Comparative descriptive characteristics of traits without their numerical expressions were mostly used within the limits of close taxa. Such descriptions referred to plant height (tall, dwarf, prostrate, rosette-shaped), habit, stem (erect, inclined, prostrate, creeping), shape of the leaf (entire, dissected, pinnate, round, elliptical, wide, narrow, lance-shaped, linear, etc.), shape of the flower (round, stellate, bell-shaped) and its organs, structure of the inflorescence, shape and size of the fruits.

Some qualitative characters cannot have numerical expressions, for example color due to anthocyanin, chlorophyll and other pigments, as well as smell and its intensity, or the taste of fruits.

Pubescence is characterized by the following qualitative traits: 1) the presence or absence of the character (whether pubescence is present or not), and moreover, whether it is due to simple or glandular hairs; 2) type of the character (position of hairs—projecting, erect, pressed, inclined, decumbent, curly; degree of hardness—hard, soft). Besides, pubescence is characterized by quantitative values expressed in such words as density and length of pubescence.

Qualitative traits of the leaf are: wrinkles due to venation, viscosity, luster. In potato species, numerical expressions of such minor traits required anatomical methods (with large magnifications) which were worked out by M.A. Sizova [8].

The importance of ontogenesis in systematics. The following are the categories of characters whose ontogenetic importance must be known to a taxonomist: qualitative (descriptive) and quantitative (numerical). Characters not requiring numerical expressions are: changing shape of the leaf from entire (in lower leaves) to dissected (in the upper ones), from wide lobes to narrow lobes; presence or absence of pubescence on different internodes, as well as the shattering of berries and duration of non-shattering nature of berries. All these characters are descriptive.

The numerical expression of the character is necessary to determine the degree of dissection as per the number of lobes and leaflets depending upon the canopy structure of the plant, i.e. the location of the leaf on the stem as per ontogenetical sequence. Ontogenesis requires prolonged study of the characters or availability of corresponding herbarium with all phases of development of the plant which is exceedingly rare in herbarial specimens of potato species, particularly those grown from seeds. Ontogenesis cannot be described on the basis of saplings of varieties because they do not reproduce varieties which have only vegetative propagation. In other words, the sapling of any variety of the potato is not the reproduction of the parent variety.

The variety of characters can be individual—on one plant, varietal or clonal (on a number of plants). The range of the variability of characters depends upon the size of the tubers, the topography of the plots, climatic conditions, geographical conditions (photoperiod, etc.).

In the previous descriptions of wild species of the potato, particularly those known to be found in only one location, even the numerical expressions of characters describe a variety of differences in a single individual plant (in the unspecified period of ontogenesis), when the herbarial material is represented by flowering branches of unknown order and even by axillary branches, but not by sympodial branches. In particular, the descriptions of the height (length) of the stem are inaccurate as they are often restricted to the phase of first flowering. This lacuna is found in the improved contemporary accounts given by Bitter, Yuzepchuk, Correll, Hawkes, Ochoa, Vargas and other scientists.

The variability of the characters of a single potato plant can be caused by the number of stems grown from one or many eyes (buds) of the tuber. In wild species, despite the presence of a number of buds on the tuber, only one sprouts by virtue of its apical dominance. Exceptions to this are the genetically initial varieties such as *S. molinae* Juz. and *S. leptostigma* Juz., in which comparatively large tubers produce many stems, generally with different levels of strength and growth. This is the reason for differences in the variety of characters on different stems of a single plant.

Irrespective of the numerical expressions for the variety of characters, many years' observations provide us with the base to characterize properly the characters of potato.

In the modern taxonomy of the potato, the statistical method of characterizing traits has not been followed by any one, because it does not help in distinguishing one variety from another.

Therefore, the establishment of many minor classes in the numerical expression of characters is conventional and reliable only for the extreme classes. But for the intermediate ones it is more appropriate to establish the range of variability, excluding overlapping into the neighboring or mixed classes. Plants as represented in a herbarium can give considerable deviations from the typical species because of ecological conditions. A classical example in this context is the description by Bitter of etiolated plants of *S. acaule* as plants of a species. It is necessary to have a reasonable number of plants grown under different ecological conditions in order to furnish a reliable description of a species. Some potato species grow very tall under poor light conditions (under the canopy of trees) in comparison with those grown in open places.

Quantitative characters, as a rule, are expressed in absolute values. The characteristics of quantitative characters are more significant in their ratio, for example, length of internodes on the stem to stem length, length of leaf to stem length, length of leaf to its width, radius of corolla to width of its petals and so on. The obtained relative values give an idea of the correlation, for example, between stem length and length of the peduncle and so on.

Equally important are the correlations among qualitative characters, particularly in the distribution of anthocyanin coloration in different organs, such as the presence of color in segments of the peduncle and other plant organs, the presence of anthocyanin in the peridermis, cortex or parenchyma of the tuber and in other organs of the plant, the correlation between a blue or red tinge of anthocyanin.

It is important to explain the correlation between the distribution of pubescence over different plant organs as well as the correlation between simple and glandular pubescence.

THE TAXA OF CULTIVATED POTATOES

Species and cultivars. In contemporary science we do not find opinion unanimous on the concept of the species of cultivated potatoes. After the discovery of the diversity of cultivated species of the potato in its native place, viz. Latin America, by the expeditions of S.M. Bukasov and S.V. Yuzepchuk (1925–1928), Linnaeus's species *Solanum tuberosum* was divided into 16 species on the basis of morphological, cytological, physiological and geographical characters. In the beginning these species were recognized by the English botanist Hawkes, also.

Later Hawkes [25] combined some diploid species on the basis of a single character, viz. absence of the dormancy period of tubers. Similarly, he combined the triploid species (except *S. juzepczukii*) on the basis of a single character, viz. chromosome number (2n = 36), ignoring significant morphological and geographical differences in these species. Both tetraploid species *S. tuberosum* and *S. andigenum* were combined by Hawkes into one species with two geographic subspecies: maritime Chilean subsp. *tuberosum* and hilly Andean subsp. *andigenum*.

Neither Hawkes nor later Dodds [21] questioned the validity of the independent status of the two species of triploid *S. juzepczukii* Buk. and pentaploid *S. curtilobum* Juz. et Buk., which had earlier been transferred by us and by Yuzepchuk to an independent series *Subacaulia* on the basis of their resemblance to the wild species of the series *Acaulia* Juz.

Other cultivated Andean species were combined by us in the series *Andigena* Buk. The wild species of potato, which seem to be the originators of the cultivated species of the series *Andigena*, are affiliated to the series *Transaequatorialia* Buk. and may also be affiliated to species of other series from the cisequatorial part of South America.

Relatives of the Chilean cultivated forms of *S. tuberosum* originate from the wild Chilean species of the series *tuberosa* (Rydb.) Buk., which are totally different geographically and morphologically from the high altitude hilly series *Transaequatorialia*.

Hawkes also combined in the single species *S. tuberosum*, not only the cultivated and wild species of Chile, but also the high altitude *S. andigenum* as a subspecies of *S. tuberosum*. Hawkes believed the Andean wild species of the series *Transaequatorialia* to be the originator of *S. tuberosum*.

Even if we admit Hawkes' hypothesis on the origin of Chilean cultivated forms of *S. tuberosum* from *S. andigenum*, which was supported by N.W. Simmonds [35], it still remains to be explained which species was the originator of the wild *Tuberosa* (*S. molinae* and *S. leptostigma*).

This certainly proves the presence of two independent origins: first, the high altitude cultivated species *S. andigenum* as well as other cultivated species of the series *Andigena*, and second, of the Chilean *Tuberosa*, the cultivated as well as the wild species.

In recent years Dodds [21] has gone further in analyzing not only cultivated *S. tuberosum*, but also all other cultivated species except *S. juzepczukii* and *S. curtilobum*, considering them cultivars of only one species, *S. tuberosum* with the following cultivars: *Tuberosa*, *Andigena*, *Stenotomum* and *Phureja*.

Each of the above-mentioned cultivars of Dodds comprise thousands of clearly distinguishable forms or groups of varieties. It is not clear as to which taxonomic units this scientist affiliates the varieties of Europe, North America and other countries, which are cultivars in the strict sense of international nomenclature, according to which the groups of morphologically close varieties should be combined in varietal types or concultivars.

It is more difficult to use these terms with regard to the endemic forms of cultivated potato species of South America. In the context of endemic local varieties from Chile, the term cultivar (variety) can be used only for sterile varieties, which are propagated only through tubers, such as Coraila and others. But fertile forms generally propagating through self-sown seedlings represent mixtures which are genetically close but morphologically different. Hence, they should be termed varietal types (concultivars) and not cultivars (varieties). This nomenclature must be applied to all forms of cultivated species of the series *Andigena* as well as *Subacaulia*.

The cultivars must be given local geographical names or names based on local dialects; only in their absence may they be Spanish, i.e. in the modern language of these countries used in scientific literature. In this context it is necessary to maintain a date-wise priority of the publication of the names of cultivars (C. Gay) [23], R. Maldonado [28], A. Canas [16] particularly in light of the descriptions given by a number of authors (Bukasov [1], A. Castronovo [17], H. Brücher [15] for Chile; W. Cevallos [18] for Bolivia; F. Herrera [26] for Peru; Yuzepchuk and Bukasov [10] for the Andean countries of South America; V. Soukup [36], Vargas [38, 39] for Peru; Hawkes [24, 25] for the countries of South America; Ochoa [31, 34] for Peru; Montaldo [30] for all countries of Latin America.

All varieties with the distinct characters of Chilean *S. tuberosum*, originated from materials of Chilean and Andean origin and introduced into Europe and South America, are affiliated by us to var. *Europaenum*.

A comparison of the classification *Solanum tuberosum* L. sens. spont.—*S. leptostigma* Juz. and *S. tuberosum* L. sens. cult.—*S. tuberosum* subsp. *chilotanum* Buk. et Lechn. = *S. chilotanum* Hawk. shows that the cultivated Chilean *S. chilotanum* Hawk. (*S. tuberosum* subsp. *chilotanum*) is distinguished from the wild *S. tuberosum* only by the diversity in the shape, size and color of the tubers and the shorter length of the stolons.

Diversity in the shape and color of tubers and flowers of cultivated *S. tuberosum* is a result of the fact that during the formation of cultivated *S.*

chilotanum natural hybridization of the wild species took place partly with *S. molinae* Juz., which has intense blue-violet coloration of tubers and flowers.

There can be no doubt that the Chilean cultivated *S. tuberosum*, as well as many man-made improved varieties of Europe and North America, originated from the Chilean wild species of series *Tuberosa* and are not linked with any wild or cultivated species of the hilly region of the Andes in South America. Hence, they cannot be considered as cultivars of the complex *S. tuberosum*, as created artificially by Dodds, in which this scientist includes different species originating from many wild species which are quite distant from the Chilean *S. tuberosum*.

Not a single known wild species of the high altitude zone of the Andes has a morphological affinity with *S. tuberosum*. Hence, the evolution of this species, with a geographic range far from that of the high altitude species and varying by 2000–4000 meters, occurred in an isolated way.

CULTIVATED SPECIES OF THE SERIES ANDIGENA

The center of diversity of Bolivian cultivated potato species is located at an altitude of 2000 to 4500 meters above mean sea level in its high altitude zone. *S. phureja* Juz. et Buk. and *S. ajanhuiri* Juz. et Buk. are endemic to this location. One more diploid species, *S. stenotomum* Juz. et Buk., is cultivated here as well as the triploid species *S. tenuifilamentum* Juz. et Buk. and *S. chaucha* Juz. et Buk., and the tetraploid species *S. andigenum* Juz. et Buk.; among cultivated species there are the triploid *S. juzepczukii* and the pentaploid *S. curtilobum*—in all not less than 8 species.

Outside Bolivia six of the above-mentioned species are cultivated in the Andes, i.e. excluding the two Bolivian endemic species. Diploid *S. goniocalyx* Juz. et Buk. and triploids *S. mammilliferum* Juz. et Buk., *S. chocclo* Buk. et Lechn. are endemic to Peru; similarly in Ecuador diploid *S. canarense* Buk. and triploid *S. cuencanum* Buk. are endemic, while diploid *S. rybinii* Juz. et Buk. is found up to Colombia.

The wild progenitors of individual cultivated species and their phylogenetic links. *S. phureja* is cultivated in a narrow geographic range which is confined to Bolivia and the lower zone of potato cultivation in the Andes (2000–2750 m above mean sea level). This zone is called "Cabezera de los valles", i.e. the upper part of forest valleys in the eastern inclines with a warm humid climate. North of Bolivia the name Phureja was not found either by Yuzepchuk in Peru, by Kessel'brenner in Ecuador or by us anywhere in Colombia.

Hawkes [24] also does not mention the name Phureja in any of the South American countries north of Bolivia. The endemicity of Phureja to Bolivia has been confirmed by the many forms of this species described in 1914 by Tsevallos.

Among the wild species of potato, *S. candolleanum* Berth. is much closer to *S. phureja* in regard to the place of its occurrence, i.e. near Sorata in Bolivia at an altitude of 2700–3500 m above mean sea level, and due to its highly dissected leaves with narrow lobes.

The above data show that *S. phureja* is a species phylogenetically distant from *S. tuberosum* and, therefore, there is no basis to consider it as a cultivar of *S. tuberosum* as done by Dodds.

Hawkes committed a second mistake by applying the name Phureja not only to Bolivian forms of potato belonging to species *S. phureja*, but also to the northern species *S. rybinii* on the basis of only one physiological character, viz. absence of a dormancy period in the tubers. Therefore Dodds' use of this Aimarian name for the northern group of species, which is also without a dormancy period, is absolutely without basis. This name relates to the Bolivian species *S. phureja* in which the tubers have no dormancy period. The northern group is represented by *S. rybinii* and other species which are not found further south of Ecuador (*S. canarense*). This is also not correct because of the fact that morphologically, and according to karyotype, *S. phureja* can be clearly distinguished from the northern species; apart from this, the introduction of Aimarian varieties far to the north, in the region of Chibcha, cultivation is less likely.

A related autotriploid variety *S. chaucha*, also without a dormancy period in the tubers, is cultivated under the name Chaucha north of Bolivia in Peru. The name Chaucha is found by E. Kessel'brenner in Ecuador also, but with the addition of the word "chuga", i.e. Chaucha chuga, belonging to *S. cuencanum*, which is completely separate from all other species, and which is a triploid species of hybrid origin from *S. andigenum* as judged by its sexually produced progeny.

Hawkes and Dodds' inclusion of all triploid species in the cultivar "Chaucha" merely on the basis of a single character, viz. triploidy, is not well founded because triploid cultivated species first of all are represented by autotriploid species (*S. chaucha*) as well as by hybridogenic ones, and, secondly, all triploid species differ considerably from each other morphologically.

On the basis of data on nomenclature as presented by Dodds in the "Classification of cultivated potatoes" (according to Correll, 1962), the following changes are necessary. The name Phureja must be attributed only to Bolivian diploid forms described as species *S. phureja* Juz. et Buk. For diploid varieties from Ecuador to Venezuela belonging to species *S. rybinii* the name Phureja must be withdrawn.

The name "Papa amarilla" (Kechuan—"Runtu papa", egg potato) used for *S. goniocalyx* from time immemorial should not be used for the cisequatorial yellow tuberous varieties as done by Dodds.

In order to avoid confusion the most appropriate decision regarding the use of the name "Papa amarilla" would be to reserve it exclusively for *S. goniocalyx* in accordance with the Indian Kechuan name "Runtu papa". Up until now "Papa amarilla" has been widely used throughout South America from Chile to Venezuela for any species or variety of the potato with yellow tubers or even some without yellow tubers.

The Northern group of South American species without a dormancy period of the tubers, Consp. S. rybinii Juz. et Buk. Establishment of a wild progenitor species for this group of cultivated species is much more complicated than for other cultivated species. In their geographic range the diversity of wild species is much less than it is further south of the equator; phylogenetically they are distant from other wild species and partly belong to the endemic series.

The most widely manifested morphological peculiarities of consp. *S. rybinii* are short stature because of short internodes, small leaves with short petioles and a small number of pairs of lobes. There are 4–6 pairs of lobes, and they are fairly wide on a short rachis.

The geographic range of *S. rybinii* sp. coll., including *S. kesselbrenneri*, extends from Colombia to Ecuador. Progenitors of *S. rybinii* might have been among the wild species growing north of the equator.

S. regularifolium Corr. from Ecuador is characterized by less vigor and strength, but nevertheless it is distinguished from *S. rybinii* by dense pubescence and very narrow leaf lobes. Tubers measure up to 3 cm in diameter. It grows at an altitude of 2150 meters above mean sea level.

It is possible that one of the progenitors of *S. rybinii* sp. coll. or *S. canarense* is more similar to *S. pascoense* in respect of its small leaves and small calyx with short mucrones and weak pubescence. *S. pascoense* Ochoa is from places further south, particularly from Pakso in Peru (Paucartambo).

Regarding leaves and other characters, *S. solisii* Hawk. (from Kan'yar, Ecuador) is similar to *S. canarense*. Conical berries are found in some diploid *Andigena* which indicates the possibility of their affinity with some *Conicibaccata*.

Besides *S. regularifolium* from the series *Transaeguatorialia* the other species resembling *S. rybinii* is *S. paucijugum* Bitt. of series *Conicibaccata* from Ecuador, which has tubers up to 3 cm in diameter.

Overlapping of the geographical range of consp. *S. rybinii* is mainly with *S. flahaultii* Bitt. which was found in Colombia in the department of Boiaka and in the suburbs of Bogota (2700 meters above m.s.l.), in the department of Kauka (3250 meters above m.s.l.) and in the departments of Santander and Narin'yu near Pasto. Its local name is "papa antigua," i.e. ancient potato.

Closer to this species is the equatorial species *S. paucijugum* Bitt. (3200–4100 meters above m.s.l.), which has edible tubers as well as big berries the size of a pigeon egg.

The name Chaucha (early), as indicated by Hawkes, is frequently used in central and southern Colombia for the diploid species, hence for *S. rybinii*, too.

As we advance from the north of South America towards the center of diversity of wild species of the potato (Bolivia with the adjoining part of Peru), we notice that fragmentation of the leaves of the diploid wild species as well as cultivated species of the potato increases. This makes it easy to establish the origin of the latter.

S. goniocalyx Juz. et Buk. was collected by Yuzepchuk in central Peru. Its Spanish name is "Papa amarilla" (yellow potato). S.V. Yuzepchuk [10] considers a central Peruvian wild species from Kanta, *S. multiinterruptum* Bitt., to be the progenitor of *S. goniocalyx*. Apart from their geographic ranges, these two species are very similar to each other in regard to high fragmentation of the leaves. *S. multiinterruptum* has up to six pairs of main lobes and numerous intermediate lobes—more than eight pairs; in *S. goniocalyx* fragmentation of leaves is high due to the large number of intermediate lobes, which is sometime more than 14 pairs. These lobes are large in both species.

The greatest geographic range and greatest diversity of forms among diploid cultivated species is observed in *S. stenotomum* Juz. et Buk. which is cultivated from Bolivia to central Peru. Hawkes [24] described 37 botanic varieties and forms of this species in large numbers from Bolivia. Only a few forms of *S. stenotomum* are found in north Peru (department Junin). Moreover, the above-mentioned geographic range is a high altitude one (usually 3000–4000 meters above m.s.l.) and highly fragmented narrow lobed leaves make it easier to establish the wild relatives of *S. stenotomum*. This species has up to eight pairs of lobes and there may be more than 20 pairs of intermediate lobes. The width of the lobes is 3–4 times less than their length.

From among wild species of the same geographic range we can exclude the rosette-shaped or semi-rosette shaped species *S. bukasovii* Juz., *S. lobbianum* Bitt. and *S. aracc-papa* Juz. and others with exceptionally narrow lobes, and those without intermediate lobules such as *S. vidaurrei* Card.

Much closer to *S. stenotomum* are the Bolivian *S. leptophyes* Bitt., *S. soukupii* Hawk. from Puno and possibly *S. candolleanum* Berth and also *S. brevicaule* Bitt.

It would have been improper to negate the possibility of hybridization between at least the above-listed species, which do not show all their diversity and variety in the emergence of *S. stenotomum*.

At the same time there is no basis to consider a unique species *S. vernei* Bitt. et Wittm., as proposed by Brücher, as the progenitor of all these species right up to *S. andigenum*.

It is equally unfounded to consider *S. immite* Dun. from the series *Tuberosa* Buk. as one of the progenitors of cultivated potato, as it is strikingly different from all *Andigena*.

In the above text we have analyzed the origin of *S. stenotomum* which was described by Yuzepchuk and by us. But *S. stenotomum*, as may be judged from the numerous forms described by Hawkes [24] and also from our observations of the collection at the VIR (All-Union Institute of Plant Industry), is represented by forms which are different from the typical form, particularly in regard to the width of lobes.

Therefore, in the origin of the wider lobed forms of *S. stenotomum* some species with wider leaf lobes as well as those with less fragmented leaves, for example *S. sparsipilum* (Bitt.) Juz. et Buk., might have participated.

Rosette-shaped species *S. bukasovii* and others could have also participated in the origin of *S. stenotomum* during their hybridization with other tall species. In fact young plants of some forms of *S. stenotomum* have very short internodes of the semi-rosette type.

The origin of S. andigenum. The diversity of the diploid cultural species of the potato provides a basis on which to presume that through natural polyploidy they gave birth to the totraploid cultivated species *S. andigenum.* No doubt, the origin of this species is polyphyletic and there is no basis to exclude the possibility of the origin of some forms of *S. andigenum* through hybridization of diploid wild species with cultivated species and even in some individual cases from tetraploid wild species such as *S. sucrense* Hawk. As a type of *S. andigenum*, Yuzepchuk and we have described a specimen of this species with highly dissected leaves, which made it possible to emphasize the considerable morphological differences between *S. andigenum* and *S. tuberosum.*

Along with this type of *S. andigenum*, which has a geographic range covering that of all diploid *Andigena* and an even somewhat wider range towards the north (up to Colombia) and towards the south (up to Argentina), there are forms of this species with less dissected or with much wider lobed leaves. The latter forms are more common towards the north, in Ecuador and Colombia (var. *tocanum*). Here the presence of diploid species of *Andigena* with much wider leaf lobes gives us a basis on which to assume that these cultivated diploid species (*S. rybinii*) gave rise to the Colombian-Ecuadoran *S. andigenum* with its wide leaf lobes.

In these countries, we observe wild tetraploid species of other series: *Conicibaccata* and *Ingaefolia*, along with the diploid series *Transaequatorialia.* Could these species be the origin of cultivated *S. andigenum*? Wild species such as *S. colombianum* Dun. and *S. tuquerense* Hawk. are highly unique in regard to all their characters; they hardly develop tubers and there is no basis for enlisting them as possible progenitors of *S. andigenum.* Moreover, triploid wild and cultivated species are not found in their geographic range. Hence, through hybridization, they did not give rise to cultivated *Andigena.*

A contrasting example is presented by tetraploid *Acaulia* which gave rise to hybridogenic triploid *S. juzepczukii* and, through a more complicated process, to pentaploid *S. curtilobum.*

The origin of triploid cultivated species of the series Andigena. The potatoes cultivated by the Indians of South America generally represent mixtures not only of forms of one species, but often also of different species in one field if these species do not significantly vary regarding length of vegetative period, dormancy period of tubers and their special purpose for making Chuno. Therefore, a mixture of tetraploid *S. andigenum* and diploid *S. stenotomum* is quite common.

Quite often natural hybridization takes place between these species resulting in triploid species such as *S. mammilliferum* Juz. et Buk., *S. chocclo* Buk. et Lechn. and *S. tenuifilamentum* Juz. et Buk. Externally it is difficult to distinguish these species from *S. andigenum* and even in their sexual progeny *S. andigenum* is dominant and the characters of the diploid parent do not become manifest. Therefore, only in the case of an extreme manifestation of an individual character may we discover which particular form of *S. stenotonum* was the diploid parent of the triploid species. Such a case is presented by *S. tenuifilamentum* which has exceptionally thin filaments; *S. stenotomum* var. *ccami* Buk. (non Hawkes) also has extremely thin filaments.

Natural hybridization of wild diploid species with cultivated tetraploid *S. andigenum* does not occur to our knowledge. This is obviously explained by the fact that such species are generally affiliated to different series, i.e. they are phylogenetically distant. A difference in the maturity period of *S. stenotomum*, which basically develops and flowers earlier than the majority of *S. andigenum*, is not a hindrance for natural hybridization because of the multistage, almost continuous flowering of *S. stenotomum* matching the flowering periods of *S. andigenum*.

The conditions for natural hybridization of the unique species without a dormancy period, *S. phureja* and *S. rybinii* grown in other periods and generally in much lower zones than *S. andigenum*, are not favorable for natural hybridization between these diploid species and *S. andigenum*.

Possibly an exception is found in the Ecuadorian triploid *S. cuencanum* Buk. which has no dormancy period of the tubers. Obviously, it is a hybrid between *S. rybinii* and *S. andigenum* with domination in F_1 (*S. andigenum* × *S. rybinii*) of the absence of the dormancy period of tubers, but in F_2 with segregation into only *S. andigenum* characteristics as in the progeny of all the above-mentioned hybridogenic species.

The fact that the high altitude geographic ranges and periods of cultivation of *S. rybinii* and *S. andigenum* in Ecuador and Colombia do not coincide was the reason for the absence of hybridogenic cultivated species in these places. In contrast to this, the species without a dormancy period—northern *S. rybinii* as well as southern *S. phureja*—gave rise to autotriploid species which are morphologically closer to the initial diploids and distant from *S. andigenum*. The sexual progeny of such species, for example *S. chaucha* (autotriploid of *S. phureja*), comprises diploid plants of the *S. phureja* type.

Autotriploids are also found in *S. goniocalyx* which is a particularly valuable species with an intense yellow pith of the tubers as in "Papa amarilla". The absence of hybrids of *S. andigenum* and *S. goniocalyx* could be the consequence of growing these species in distant fields. It is also worth mentioning that this absence of hybridization is confined to the northern border of the geographic range for the southern diploid species of the upper zone represented by *S. stenotomum* and *S. goniocalyx*.

A third species of this zone, frost-resistant *S. ajanhuiri*, does not have autotriploids. This may be one more indication of its affinity to wild species as was mentioned above.

S. stenotomum gave rise to two types of triploids. Apart from the above-mentioned hybridogenic species *S. angidenum* × *S. stenotomum*, which domination of the first component, other autotriploids of *S. stenotomum* are also known.

The triploid species *S. mammilliferum* is common in Puno at an altitude of 3400–3700 and 3840–4300 meters above m.s.l. under the names Soccohuaccotto, Chackchepfuru, Mocketa, Cketosa, Machacc-sencco and others. It yields more than *S. stenotomum* and is more resistant to fungal diseases and pests.

S. *tenuifilamentum* is cultivated at an altitude of 3650–3850, sometimes up to 4300 meters above m.s.l. and *S. juzepczukii*—at 3600–4400 m. Local names in Bolivia are: Kaisalla, Ruqui, Orco-ruqui, Malccu-aparu and Luqui. There is no doubt that the triploid species of the series *Andigena* are not limited to the above-listed species. Discovery of new species of this series requires counting of the chromosomes of all specimens of any collection, except for diploid species which may be identified even by external characters, particularly on the basis of poor pubescence in the upper surface of the acumens of the corolla. Ochoa [32] in his collection of 518 specimens collected in the basin of Lake Titicaca in Bolivia and Peru, established chromosome numbers for 469 specimens which were comprised of tetraploids 170—*S. andigenum*, pentaploids 19—*S. curtilobum*; diploids 125—*S. stenotomum*, 12—*S. goniocalyx*, 18—*S. ajanhuiri* and 4—*S. phureja*. In all there were 159 diploids, i.e. about 35%. Some 111 triploids, including 49—*S. juzepczukii*, comprised 22%. The remaining 62 triploids were not affiliated to the species described earlier by us, and no doubt there are unique forms among them.

From the above-mentioned data of Ochoa it is evident that at the center of diversity of cultivated potatoes, in the basin of Lake Titicaca, *S. andigenum* comprises only about 1/3 of the specimens—almost at par with diploid species. Here the frequency of hybridization with *S. andigenum* is extremely high because every two to three specimens of diploids gave rise to one triploid. Similarly, hybridization of diploids with *S. acaule* is also often observed.

As we move away from this center of origin of cultivated potatoes, the frequency of emergence of triploids drastically declines and among 385 specimens collected in northern Peru (north of Lima), Ochoa [31] discovered only one triploid. It was not unexpected that among 385 specimens 24 were given the name "Chaucha" (beyond the limits of the geographic range of the Bolivian species *S. phureja*), and no doubt they belong to the Ecuadoran forms of *S. kesselbrenneri*. Thus the local name "Chaucha" is not confined to triploid derivative species of *S. phureja*, particularly to the species described as *S. chaucha* (*S. surimana*).

Therefore in botanical nomenclature which does not carry a definition as per morphological characters and when identification with *S. chaucha* Juz. et Buk. is not supported by the chromosome number, it is more appropriate to use the name *S. surimana*.

The following changes are necessary in the nomenclature of cultivated species of the potato.

Cultivars comprising diploid consp. *S. rybinii* have the local name Chaucha which must be retained with them. In order to avoid confusion with triploid *S. chaucha* Juz. et Buk., which was wrongly expanded by Hawkes to all triploid cultivated species, it is worthwhile to abolish the names *S. chaucha* Juz. et Buk., *S.* × *chaucha* and groups of triploid cultivars "Chaucha". The name Chaucha should be retained only for diploid cultivars of *S. rybinii* because it is widely used in Colombia and Ecuador for corresponding cultivars.

All cultivated triploid cultivars emerging spontaneously from the hybridization of *S. andigenum* (4n) with a diploid species and not clearly distinguishable from *S. andigenum* (except *S. tenuifilamentum* and others) should be designated as *S. andigenum* (3n), but as *S.* × *chaucha*, as suggested by Hawkes.

Judging partly by the many local names in different countries and their departments in South America, the most numerous cultivars are of *S. andigenum*. Even with the exclusion of synonyms, the number of cultivars of *S. andigenum* numbered in the hundreds and may be in the thousands.

The number of known cultivars of the diploid species *S. stenotomum* and the consp. *S. rybinii* is probably about 100 in each of these species.

The number of cultivars of endemic Bolivian diploids *S. phureja* and *S. ajanhuiri* is small and corresponds to variants of the anthocyanin coloration of their tubers, i.e. it is less than ten. Less diverse are the cultivars of triploid species, as represented in one or two variants according to anthocyanin coloration. Such cultivars are *S. surimana* (*S. chaucha*), *S. cuencanum*, *S. tenuifilamentum*, *S. mammilliferum*, *S. chocclo* and *S. juzepczukii*, as well as pentaploid *S. curtilobum*.

It is very difficult to count the cultivars of *S. andigenum* (3n) because they are morphologically indistinguishable from *S. andigenum* (4n). Maybe their sterility should be considered as a criterion. The number of such triploid cultivars corresponds to the frequency of spontaneous hybridization and in Bolivia and in Southern Peru they number in the hundreds [2, 22, 25, 33].

According to V.N. Sinel'nikova [9] the number of the names of cultivars in Chile is about 200. Among them are native synonyms as well as names specially translated into the Spanish language. This is well known for the cultivars of *Andigena*. The term "cultivar" must be used in a somewhat wider sense for local Indian potatoes than for the man-made improved varieties (clones) because they can be a mixture of seedlings (populations from natural self-pollination). The number of cultivars of man-made improved varieties is particularly high, and reaches a few thousand.

The large number of cultivars makes it necessary to combine them into groups of different categories. These groups must be created in conformity with the evolution of the cultivars.

During cultivation of the wild species selection was based on yield potential, i.e. on the boldness of the tubers and then on their shape and color. The general appearance of the plant, as shaped by its foliage, was a lesser basis for selection during the initial stages of cultivation of this crop. Therefore, the general appearance of the plants, i.e. their leaves, must be the basis for creating groups of category I. Further division of these groups of category I into groups of category II must be done according to the morphological characters of the flowers which were not subjected to artificial selection. Groups of category III are distinguished by the distribution and tinges of anthocyanin on plants that emerged naturally—through mutation as well as through artificial selection according to the color of the tubers. Groups of category IV, based on the shape of the tubers and the color of their pith, are the result of artificial selection.

These same principles can be applied to the groups of man-made improved varieties. The complexity in the groups of these varieties lies in the fact that along with the varieties of *S. tuberosum* originating from the Chilean varieties (*S. chilotanum* Hawk.), there are varieties such as var. × *europaeum* during whose development hybridization of Chilean *S. chilotanum* with *S. andigenum* occurred.

During the past few years a very large group of varieties has emerged, viz. interspecific hybrids with different wild species of potatoes. Phylogenetic principles are required for their grouping into varietal types and so on.

Cultivation of the potato by Indians in the basin of Lake Titicaca results in a mixture of different forms of *S. andigenum* and *S. stenotomum*. Besides natural hybridization between these species, and hence the formation of triploids of generally shorter life due to their greater susceptibility to viruses, this can result in the evolution of diploids of *S. andigenum* and also tetraploid hybrids with participation of the non-reduced gamete of the diploid species. Therefore, this is the third way that tetraploids can emerge, besides self-pollination of *S. andigenum* and hybridization of different forms of this species. Variations in the local cultivars are restricted by the geographical isolation of the numerous hilly valleys, but exact data on the degree of diversity of cultivars of *S. andigenum* or *S. stenotomum* in these isolated regions are not yet available.

Dispersal of cultivars between regions is possible through birds feeding on the potato berries. In *S. andigenum* evidently self-fertilization is not uncommon, and this restricts slightly the variability in self-grown plants.

Polymorphism of hybridogenic triploid cultivars is governed by the polymorphism of the parental species. It is worthwhile to mention as an example the slight diversity in cultivars of triploid *S. juzepczukii* as a result of the slight diversity in the forms of the wild tetraploid parental species *S. acaule*. Exclusive domination of the majority of characters of *S. acaule* in *S. juzepczukii* suppresses the manifestation of characters of the diploid parent *S. stenotomum* which is, no doubt, endowed with much richer forms than *S. acaule*.

Therefore, *S. juzepczukii* is represented by less polymorphic cultivars which continue to be recreated continuously with the death of much older clones due to viruses. According to Wight [43], type *S. juzepczukii* was mentioned as early as 1613 by Bertonio under the name "Luque".

Among the unsolved problems relating to the evolution of South American cultivated potatoes, there is a question as to the importance of self-grown plants through seeds in the origin of cultivars. The presence of numerous triploid cultivars proves the frequency of natural hybridization of *S. andigenum* × *S. stenotomum*. From such hybrid seeds triploid cultivars emerged from self-sowing, because generally the Indians did not collect and did not sow the potato seeds. There is no doubt that self-sowing simultaneously gave rise to cultivars through self-pollination as well as through natural hybridization of *S. andigenum* and other species.

It is particularly interesting to establish how frequently the formation of hybrid berries on *S. acaule* occurs in its zone adjoining the geographic range of *S. stenotomum*. What percentage of triploid *S. juzepczukii* grows in a self-sown population among plants of *S. acaule*? How is the selection made by man in the mixture of plants of *S. acaule* and *S. juzepczukii*? Obviously, according to the yield of the tubers. How long is it possible to vegetatively propagate *S. juzepczukii* in its native place? Under our conditions *S. juzepczukii* quickly perishes due to viruses and even original specimens become susceptible to viruses.

The tetraploid *S. acaule* is probably the only species from the selection of *Tuberarium* with 100% self-fertilization. *S. acaule* var. *albicans* has white flowers and 72 chromosomes and is characterized by low self-fertility. Self-pollination in *S. acaule* takes place during the very early stages of flowering, may be even in the bud as is sometimes inferred from futile emasculation of flowers. Therefore, natural hybridization of this species with *S. stenotomum* can take place only on the opened flowers of *S. acaule* with combined (additional) pollination by the pollen of *S. stenotomum*. This hypothesis requires experimental verification.

Other possible methods of natural hybridization of *S. acaule* with diploids are: 1) reduction of self-fertility or emergence of self-sterile forms through mutation; 2) damage to anthers in the buds by insects such as honeybees as they feed on or collect pollen.

The problem of the origin of *S. curtilobum* awaits an experimental solution through berries derived from self-pollination in isolation and a study of the first generation. Seeds of *S. curtilobum* obtained from Peru by Ochoa gave a predominantly tetraploid generation and only a few diploid plants; among both groups there were plants with characters of *S. curtilobum* as well as absolutely unique plants which were different from all species known to us. Among tetraploids there were plants of the type *S. andigenum*.

Following Hawkes, Dodds assumes that the type of *S. tuberosum* is the plant specimen described by Linnaeus and brought to Europe from the Andes of South America, hence—*S. andigenum*. In view of this, Dodds includes all cultivated potatoes of the New as well as the Old World in *S. tuberosum* L. s. l., except cultivated Subacaulia, and subdivides them into four subsequent taxonomic steps.

1. According to ploidy level. Diploids (2n), triploids (3n) and tetraploids (4n).

2. Within the limits of these categories they are divided into groups corresponding to the botanic species established by Yuzepchuk and us, but with a combination of some of them and with a change in the nomenclature.

Besides *S. stenotomum*, Dodds includes *S. ajanhuiri* also in the group *Stenotomum*. The group *Phureja* is comprised of consp. *S. rybinii* with a geographic range from Colombia to northern Peru. In fact Dodds does not include type *S. phureja* from Bolivia in his classification, but mentions that absence of the dormancy period in tubers is observed among other diploid groups, hence in the group *Stenotomum* also.

If we understand the interpretation of Dodds in this way, then Bolivian *S. phureja* Juz. et Buk. (non Hawkes) shows the morphological closeness of *S. phureja* with *S. stenotomum*. In fact, Yuzepchuk in his unpublished work "Opyt Sistematiki Kartofelya" (Experiments in the systematics of potato), combines *S. stenotomum* along with *S. phureja* in one group *Stenotomum*. This particularly emphasizes the wrong use of the specific name *S. phureja* of this endemic Bolivian species for the Colombian-Ecuadoran (consp. *S. rybinii*).

3. The third taxonomic category of Dodds is the subgroup with species *S. goniocalyx* (subgroup *Goniocalyx*). It represents a subgroup of the group *Stenotomum*.

In the group *Phureja*, Dodds obviously and unsuccessfully identified a subgroup *Amarilla*. In this subgroup he combined botanical cultivars which had a yellow pith of tubers and central Peruvian cultivars of *S. goniocalyx* with white flowers.

4. The fourth category of Dodds: cultivar-variety-clone. Hence, between a species and a cultivar he distinguishes three more intraspecific categories of

cultivated potatoes: 1) according to ploidy level; 2) according to dormancy period of the tubers (*Stenotomum* and *Phureja*); and 3) according to geographic range (*Andigena* and *Tuberosa*). Besides these there are subgroups according to color of the pith of tubers of *Amarilla* and other characters.

THE CHROMOSOME NUMBER AS A TAXONOMIC CHARACTER

The presence of autotriploids in a small number of species of the section *Tuberarium* is often not characterized by significant changes in their morphological characters. Such triploids not distinguishable from the initial diploids are known among *S. bulbocastanum*, *S. cardiophyllum*, *S. chacoense* and *S. maglia*. In this context, in *S. maglia* and others even the obligatory inherent sterility of triploids is not a trait, making it possible to distinguish these triploids from the initial diploids, because diploids are also sometimes self-sterile. Sometimes only an analysis of the pollen gives us the opportunity to distinguish triploids.

Differentiation of autotriploids from triploids emerging through hybridization of (4n) and (2n) is also sometimes difficult. For this purpose the character of absence of a dormancy period in tubers is not reliable, but this is considered by Hawkes to be typical for the unification of all diploid species without a dormancy period in tubers of *S. phureja*.

At the same time in *S.* × *chaucha* Hawkes united all triploid species established by us without a dormancy period in tubers of the species *S. chaucha* (3n) and *S. cuencanum* as well as with a dormancy period in tubers (*S. tenuifilamentum*, *S. mammilliferum*, *S. chocclo*).

Absence of a dormancy period in tubers of the hybridogenic *S. cuencanum* provides a basis for assuming the dominance of this character. Very thin filaments of hybridogenic *S. tenuifilamentum* also gives a basis for assuming dominance of the above-mentioned character. To some extent this is supported by the much thinner filaments of "dominant" *S. andigenum* in comparison with the "recessive" *S. tuberosum*.

All that has been said above about triploids does not provide a basis for considering this chromosome number ($2n = 36$) as a good taxonomic character to identify triploids in an independent taxonomic category on a par with diploids and tetraploids.

Along with Yuzepchuk we isolated triploid species according to their characteristic morphological characters. The chromosome number in them was later established by V.A. Rybin (7) as an additional, but not a basic, taxonomic character. Therefore, neither the species *S.* × *chaucha* Hawk., nor its renaming by Dodds in the group of cultivars "Chaucha", can be accepted as wrongly interpreted "volume" of the species or in the sense of a unified group of cultivars.

Dodds considers it impossible to divide endemic forms of the cultivated potato into species on the ground that they are often cultivated as a mixture of forms easily crossing with each other and having setting characters (morphological). At the same time, as an illustration of this "setting of morphological characters", he uses only seven characters and that too in a general expression which is far from being adequate to differentiate the forms.

Differentiation of species can be established: 1) according to the variety of morphological and biochemical characters; 2) according to morphology of chromosomes (karyogram, pachytene); 3) according to meiosis; 4) according to compatibility.

During the first stage the young species were self-fertile and self-compatible. Incompatibility sets in with the evolution of species as a consequence of considerable differentiation of morphological, physiological and biochemical properties of the species. Compatibility of genomes is not a basis for the unification of young species such as *Andigena* and *Transaequatorialia* (and also *Tuberosa*).

Incompatibility sets in during the later stages of evolution, i.e. in much older species (of North American subsections *Pinnatisectum* and *Cardiophyllum*).

The youngest species (*Acaulia* and *Demissa*) are easily crossed with species of the preceding stage of evolution (all diploid series of South America).

In the Andean group of species of the series *Transaequatorialia* we can differentiate species with a number of intermediate lobes between adjacent pairs of lobes of the leaf. There are more than two pairs and sometimes more than eight; when including acroscopic ones there are more than ten.

The following species of the series *Simpliciora* have been isolated from *Transaequatorialia* according to this character: *S. simplicifolium*, *S. venturii* and *S. gandarilasii*. From the stock of *S. curtilobum* and from seeds sent by Ochoa from Peru (k-3558) we got three plants each with the chromosome number 2n=24 and 2n=36. These plants were low statured with narrow lobed leaves (three intermediate lobes). This character distinguished them from all *Andigena* known to us, as well as from wild species of the section *Tuberarium*. It is possible that we will have to search for progenitors of these diploids among low-statured species which evolved through losing high segmentation of leaves as a result of loss of the intermediate lobes. It is also possible that one example of such leaf evolution is presented by *Rybiniana*, particularly *S. rybinii* s. str. in the group of wide-lobed *Andigena*.

Ancient Aimaranian names for the endemic forms of the cultivated potato presented by Bertonio show that the Indians easily distinguished species of cultivated potato which were afterwards established by Yuzepchuk and us and were also accepted by Hawkes. Such names are Luqui—*S. juzepczukii*, Chocke-pitu—*S. curtilobum*, Alcca phinu—*S. stenotomum*, Surimana—*S. phureja*. Other names, except the names of wild potato, obviously relate to *S. andigenum*.

There are two main geographic zones of potato cultivation in Colombia. One of them is the high altitude "paramo" where the varieties Paramuna, Caiceda, Tuquerena and others with narrow-lobed leaves of the species *S. andigenum*, are being cultivated.

Second is the low-lying zone of Sabana de Boyaca and Sabana de Cundinamarca where wide-lobed *S. andigenum* Sabanera, Tocana and others are cultivated.

Probably these more common varieties of *S. andigenum* were the first ones to be described by the Spanish chronologists under the Indian name Yomsa (potato) which did not remain in vogue for long, being replaced by the Kechuanian name Papa. At present here as well as in Ecuador the name Chaucha is prevalent; this is used partly for the native two-cropped diploid and triploid varieties of *S. rybinii* and others. The Aimaranian name for the cultivated potato is Amsa and that of the wild potato Afora [18].

HISTORY OF THE SYSTEMATICS OF CULTIVATED VARIETIES OF POTATOES

The systematics of cultivated varieties of the potato are represented by two historical stages. In the first stage taxonomists reviewed the so-called commercial varieties of the potato which were developed in the 19th and 20th centuries in Europe and the USA. These were the progenies of potato specimens of native varieties of Chilean and Peruvian origin which had been brought in from South America.

In our time Soviet expeditions to South America established that the Chilean native varieties belong to the species *S. chilotanum* Hawk. But the native potato varieties introduced into Europe belong to *S. chilotanum* Hawk. as well as to *S. andigenum* Juz. et Buk., and possibly even to diploid cultivated species [6].

Only the Chilean varieties developing tubers under the long-day conditions of Europe and the USA could have remained in the initial form. However, while propagating them through tubers such varieties perished due to viruses and other diseases. Only the progeny of specimens producing seeds and multiplying through seeds survived.

Native varieties of Peruvian origin, as a rule, were fertile and could be reproduced through seeds, giving sexual progeny that were also fertile but which did not develop tubers under long-day conditions. Therefore, these native varieties had a short life and within a short period of time they perished. The probable incidental hybridization of *S. tuberosum* L. with *S. andigenum* could have given varieties with the dominance of *S. tuberosum* L., not only in regard to the long photoperiod needed for tuber formation but also in regard to morphological characters. Therefore, the type *S. tuberosum* described by Linnaeus was *S. chilotanum*.

In the evolution of the cultivated potato Linnaeus's European species *S. tuberosum* appears to be the third stage and representative of the third center of its diversity emerging beyond the limits of the original location of the potato, i.e. South America. The geographic range of this center is very wide and it covers all countries cultivating *S. tuberosum* beyond the limits of South America. This means that our assumption that *S. tuberosum* carries characters of *S. andigenum*, and hence can be named *S.* × *tuberosum*, is not based on fact. Therefore, Linnaeus's initial name *S. tuberosum* L. remains in vogue but it does not cover *S. chilotanum* which has a geographic range in Chile—the second center of the cultivated potato.

The main task is to establish whether the contemporary native varieties of Chile and their sexual progeny have the characters of *S. tuberosum* which were described by Linnaeus. Rapid extermination and death of old native varieties of Chile make this task exceptionally difficult and probably impossible.

How did the systematics of varieties of the European type, including varieties of the USA, develop? The first description of potato varieties was given by Putshe in 1819. He described 33 varieties with colored pictures of tubers whose types are found in contemporary native Chilean varieties [6]. Descriptions by Alefeld [13] also give only a list of varieties, but not their classification in the desired sense. It was only Vilmorin who in 1902 gave a system of varieties according to the shape and color of the tubers. The effort to group varieties according to a complex of characters, including characters of the bush, can be attributed to K. Snell (1922) for varieties of Germany[1] and to V. Stuart (1905) for varieties of the USA. Later on similar descriptions of varieties were given by McIntosh [44] and E. Klyapp [5].

In the USSR descriptions of the collections of varieties were given by Aseeva and N.A. Lopatin. The first classification of varieties was given by Bukasov in 1923. Unfortunately, it gave numbers instead of the names of varieties because there was no guarantee as to the authenticity of the latter. It was only in 1971 that for the first time a description of varietal types was given by L.I. Kostina [6].

Simultaneous with the description and listing of potato varieties beyond the limits of their native place, viz. South America, there appeared much earlier lists of names of the native varieties, not in agronomic literature but in dictionaries, particularly in the dictionary of the Aimarian language. Bertonio reports names of more than ten native varieties. Judging from these names of native varieties in Bolivia, the Indians had distinguished varieties from ancient times; these were correlated by Soviet scientists [10] into different species including the most common ones such as *S. andigenum*, *S. stenotomum* and *S. juzepczukii*.

Verification of the species affinity of varieties reported by Bertonio is diffi-

[1]Here and elsewhere this refers to the boundaries of Germany as they existed until 1949.

cult because of the fact that some names had been given to different varieties which had some common characters. For example, the name Luqui (sometimes written as Ruqui) was given to the most frost-resistant species, *S. juzepczukii* and *S. curtilobum*. At the same time *S. juzepczukii* has its own name, Orcco-malcco, in contrast to the name China-mallko which related only to *S. curtilobum*. It is interesting to compare the use of the words "China" and "Orcco" in compound names of other varieties except those which belong to the two above-mentioned frost-resistant species. The words "China" and "Orcco" refer to female or male peculiarities of the varieties. However, it is not clear how these differences are expressed. *S. juzepczukii* is the species with sterile flowers. Maybe these words are related to other properties of the species or variety, for example boldness of the tubers or their yield potential. South American solanologists must help verify these differences.

A description of the native varieties of Bolivia, published by Tsevallos in 1914, was exceptional among South American works of a similar nature in regard to the number of listed varieties as well as in regard to the colored illustrations of the tubers.

It must be mentioned that in the work of Wight [43] the native Bolivian varieties are combined like they are in all other subsequent works that list the Bolivian varieties particularly the works of Kardenas and Ochoa.

In the Peruvian works published after our book "Kartofeli Yuzhnoi Ameriki" (Potatoes of South America) [1], the authors Soukup [36], Vargas [38, 39] and Ochoa [31–34] stick to our specific systematics and method of description in a very brief form. A more exhaustive list of the varieties of Peru was published by Ochoa. His work contains a description of about one thousand native varieties with their original names in the Kechua language and a brief description of the tubers. The chromosome numbers of each species have been counted which made it possible to establish their specific affinity.

Some native varieties of northwest Argentina have been listed by Viirsoo [41].

Alefeld [13] for the first time gave Latin and Latinized names to varieties (in the year 1886). He listed 55 varieties with their brief description. This was the final picture on the development of diversity of varieties preceding the beginning of potato breeding in Germany in association with the firm of Richter, Paulsen and Tsimbal.

At the beginning of the 19th century the number of described varieties increased to a few hundred [40]. For 50 years after the book of Putshe the number of European varieties in West Europe, excluding England, increased from 34 to 55 varieties; these were first systematized by Alefeld. Alefeld subdivided the varieties (which were described by him) into two main groups according to the shape of the tubers: 1) long, asymmetrical, often bent like a horn; 2) short, from round to kidney-shaped. A detailed description of the shape of tubers within the two above-mentioned groups was made according to the depth of the eyes (buds) and the presence of wrinkled protrusions above the eyes resembling spruce cones in their shape (Tannenzapfen in German). Varieties with flat tubers were also isolated.

Within the above groups, which were based on the shape of the tubers, there are subdivisions based on the color of the tubers and their pith. Bluish-black color was recorded in the variety "Aethiopicum" (i.e. Ethiopian) which was afterwards known as Negresse (Negro in French). Later on groups of varieties with red tubers and yellow pith were also isolated. The coloration was recorded as entire, variegated, bicolored with colorless portions in the colored background or vice versa. A combination of two anthocyanin colors (blue and red) was recorded in only one variety called "bicoloratum" (bicolored). The combination of intense black-blue with a similar entire color of the pith was recorded in a variety which was given the Latin name "tinctorium" (dyeing), thereby indicating the ancient Indian method of using the juice of such tubers to dye cloth.

As a whole the work of Alefeld was the first systematic account of contemporary varieties of potato using characters not only of the tubers, but also of the stem, leaf, inflorescence and berries. Alefeld mentions the colored pictures of Putshe which were the first to appear in literature with the varietal description of the potato. There were no works on the systematics of English, American and other varieties contemporary to the work of Alefeld.

Much later lists of varieties appeared in the catalogues of the firms Haage and Schmidt as well as in certain articles and other catalogues. The most exhaustive among them was the catalogue published by Vilmorin in 1902 [40].

Simultaneously the planned breeding work of varieties began. In Germany a series of varieties were developed by Richter which included such well-known varieties as Imperator of Richter which is widely cultivated even today in Transcaucasia (Georgia) under the local name "Asuretuli". Among other plant breeders of that time we must name Paulsen (varieties Juli and Move) and Tsimbal who developed varieties such as Woltman, Saksonskaya Lukovitsa and Russkaya Lukovitsa. It is interesting to note that the name of the last variety was also mentioned by Alefeld.

In the same period L. Burbank began potato breeding work in the USA on the basis of *S. chilotanum*. His varieties Early Rose and Burbank are still under cultivation. In England breeding work on the potato was carried out by Sutton and in Holland by Wenhoizen and others.

Much later the series of renamed new varieties, such as Magnum bonum and others, made it necessary to establish synonyms for the varieties and to form special committees on synonyms for potato varieties in varietal breeding literature. A detailed description of English varieties was given by R. Salaman in 1926 and later by McIntosh in 1953.

In the history of the systematics of potato varieties an exclusive place is occupied by the work of Klyapp [5]. No other work can surpass it even today in respect of the thoroughness of the description of varieties. It is only in methodology that the details of varieties published at the VIR (All-Union Institute of Plant Industry) can compete with the descriptions of Klyapp.

Modern Soviet as well as foreign atlases of potato varieties comprise a description of varieties registered as varieties accepted or allowed for cultivation in the respective countries, but which were released for cultivation or which are maintained from earlier years in the USSR. Such atlases exclude varieties with synonyms.

Atlases are generally appended with colored illustrations of tubers, leaves, inflorescences and sprouts. There is no herbarial documentation. Details in the descriptions in them are limited in comparison with botanical literature. This is because, first of all, these atlases are designed for use within the territories of the respective countries and rarely beyond them, if the introduced varieties are cultivated without any restriction by the country where they were developed. Hence, each character of a variety must be depicted as its average expression excluding the extreme versions caused by variations in characters due to environmental conditions. The atlases are meant for use under field conditions where it is difficult to determine minute details of the differences in pubescence or glands by a large group of seed producers and quality control specialists.

While establishing varieties with synonyms the descriptions can give more minute details than the characters described in the atlases of the varieties. The description of varieties recommended for the state varietal trials should be particularly elaborate, in order to avoid synonymics or the release of new varieties which are difficult to distinguish from each other.

Systematization of varieties developed in a certain country must give a generalization of the germplasm with the definite initial geographic ranges of the given country in the form of varietal types and relative varieties. An example of the description of varietal types is the work of L.I. Kostina [6]. In this work considerable attention is given to the description of sprouts of the tubers grown under experimental conditions and which can be easily reproduced for identification of varieties.

While identifying varieties according to the shape and color of the tubers it is necessary to take into consideration the high variability of these characters due to environmental conditions. Therefore, while describing the shapes of the tubers, Klyapp made use of the statistical method, taking measurements of hundreds of tubers. While describing the color of tubers one should use a corresponding scale of colors as given in the 1975 atlas of colors.

Outside South America improved varieties of the potato are the result of artificial selection from the complex genetic population obtained through several crossings. This created great difficulties initially for the systematics of closely related populations within the species *S. tuberosum*. In recent years interspecific hybridization has been used extensively, not only with other cultivated but also with a few wild species, particularly *S. demissum*, *S. stoloniferum*, *S. acaule* and others.

The methods of the systematics of native South American cultivated varieties are significantly different from those of the potato varieties cultivated in Europe.

The basic peculiarities of native cultivated potato varieties are their affinity to different potato species within the narrow limits of their geographic ranges, and their clear distinction according to chromosome number 2n=24, 36, 48 and 60 as established by V.A. Rybin [7]. Each of the cultivated species is clearly distinguished by phylogeny leading to different wild species and is represented by its own more or less numerous varietal assortments.

A significant distinctive peculiarity of native varieties of the potato is their antiquity. Man-made improved varieties of the potato *S. tuberosum* are an exception, and have an age of about one hundred years. The ages of the native varieties are several hundred years. The native potato emerged several thousand years B.C., as judged from drawings of the potato in the remains of ancient valley cultures.

Therefore, the actual age of the native varieties can be established only on the basis of the history of the potato as a cultivated plant, which varies morphologically depending upon the habitat in Peru or in other countries and also on the basis of the linguistic history of its varietal names.

The works of Ochoa [31, 34] are particularly valuable for the systematics of native varieties of potatoes. He collected more than one thousand varieties in Peru, often from isolated pockets which are only accessible on foot. Varieties localized in a few solitary pockets are of particular interest. This proves their greater antiquity in comparison with the varieties of smaller geographic ranges. The importance of local names for the native varieties increases because they are strictly confined to different potato species, a fact which is supported by the difference in chromosome number as presented by Ochoa.

During hybridization of the European *S. tuberosum* with cultivated varieties, we acquired knowledge of the genetic potential of not only the pairs of varieties in the crosses, but also of both the pairs of parents of these varieties and sometimes of their grand-parental varieties. In crossings with native varieties we do not know anything about their parents and, hence, nothing of their gene composition. That is why systematics of native varieties of the potato are essential to provide a basis for the study of their progeny, even during hybridization with varieties of *S. tuberosum*.

The age and geographic range have three aspects: 1) in wild species, 2) in native varieties, 3) in the man-made contemporary varieties of potatoes. The age of the wild species is calculated from the time of its discovery and first description. The geographic range of these species is insignificant and in the beginning it was confined to the original place of detection of the species. In the first detected wild species, beginning with *S. tuberosum*, the antiquity of the description of the species is calculated to be up to two hundred years, but for the majority of species it is less than one hundred years. In the case of the cultivated species discovered by the Soviet botanists it is less than fifty years.

The geographic range of the species does not depend on its age, but only on the homogeneity of ecological conditions and the degree of variegation of the relief of the area, which may depend on geographic conditions including volcanic phenomenon. The latter creates an interruption in the geographic range; for example, the original habitats of *Acaulia* are highly variegated and isolated if they are confined to different high altitude ranges of the Andes.

It is quite different to determine the age and geographic range of improved varieties that have emerged in Europe and other countries outside the limits of the native potato cultivation. The age of the varieties known from their first description in the works of Putshe (1819) is much less than two hundred years. On the other hand, our contemporary varieties have an age of about one hundred years. But the concept of a geographic range for contemporary improved varieties is not as much geographic as commercial. The varieties emerge in one of the countries of Europe, North America and other continents and their primary geographic range was confined to a small area of the fields of the plant breeder before the official registration of varieties.

In the USSR the age of new varieties is calculated from the date of their release for cultivation by the State Commission (Goskomissiya). Till that date its area is limited to a few hectare. Later the geographic range is determined only by the quickness of its dispersal and demand by the farms covering large territories, as in the case of the variety Lorch.

In other countries the geographic range of the variety expands initially within the country where the variety was developed. This range covers an area of a few hectare in the beginning, but subsequently it covers up to thousands of hectares. Some varieties become popular in neighboring countries and even in other continents.

The names of taxa to be redescribed should not repeat the taxa already published in the "Index Kewensis" with its addenda as well as "Index Londinensis" where all diagrams of species published to date are given. The names are given according to the place of growing (*S. vallis mexici*, *S. guaytecarum*), the name of the botanist who collected it for the first time (*S. commersonii*, *S. henryi*, *S. ohrondii*) or in honor of a famous botanist (*S. juzepczukii*, *S. vavilovii*) as well as on the basis of the local name of the specimen (*S. phureja*, *S. maglia*).

In the last case the correct pronunciation must be given; for example, "phureja" must be pronounced "purekha" and not "fureya". A name based on a character common for two or more taxa is not permissible. Similar mistakes are unavoidable if a given name is not accompanied by its meaning or sense in the Russian language.

Some local names are used for different varieties which may even belong to different botanic species. For example, "Amarilla" (yellow) is used for yellow varieties of *S. andigenum* as well as of *S. tuberosum*.

The addition of the Kechuan name "papa" to varieties of Chilean origin is not allowed as it was not in use in the Maluche and Chilote languages of the Araukania and Chiloe islands, respectively. The adjective constituting the name of varieties should never be given to names of botanical forms and varietal types, such as "sunchchu" (long).

Besides this, it is necessary to consider that some native names for varieties characterize a quality in them which is common for a number of varieties; for example, "Ccompis" means the good flavor qualities of a group of different varieties. There are instances when different varieties in adjacent villages have the same name. It is possible that in the pre-Inca languages such a name depicted the same concept or quality of geographically close varieties of the potato.

The morphological characters of species and their taxa are the basis for the botanical systematics of the potato. The species should never be singled out on the basis of physiological characters—for example, the dormancy period of the tubers—because these are secondary characters accompanying certain morphological characters of species.

Breaches in this principle led to the mistakes of Dodds and Hawkes who combined into one species, *S. phureja*, two species clearly distinguishable from each other in regard to morphological characters, geographic range, and phylogeny: *S. rybinii* and *S. phureja*. In fact, except for *S. phureja* and *S. rybinii*, a short dormancy period of the tubers is present in *S. stenotomum*, *S. andigenum* and some varieties of *S. tuberosum*.

Apart from this, separation of species on the basis of chromosome number without any consideration of morphological characters is also a breach of the principles of botanical systematics of the potato. An example of such a breach is the separation by Dodds and Hawkes of a united, cultivated species of potato, *S. chaucha*, on the basis of chromosome number ($2n = 36$). In fact, the *S. chaucha* these scientists refer to includes many species clearly distinguishable according to their morphology, geographic ranges and phylogeny: *S. tenuifilamentum* and *S. mammilliferum*, as well as all spontaneous triploids of *S. andigenum* which are morphologically indistinguishable from tetraploid forms of *S. andigenum*.

Latin or Greek names, with their translation into modern languages [14] including Russian, are accepted in the nomenclature of the taxa of wild potato species.

Names for the cultivated varieties are given in the language of the country where the variety was developed, with the addition of the name of the breeder or the firm that developed the variety. Victoria of Paterson, Imperator of Richter and many others are examples of such names.

Sometimes a series of varieties developed at one center or institution is given a common adjective. Such is the series of varieties of the Scotch breeding station at Pentland: Pentland Ace, Pentland Crown and others. The series of the Irish varieties is unified by the name Ulster: Ulster Premier and others. Groups of varieties from Colombia are released with the word Diacol.

The great demand for new varieties in England during the end of the 19th and beginning of the 20th century caused a wave of misuse of potato varieties in the trade. This led to a situation where the number of synonyms or similar names increased to hundreds for the most popular varieties such as Magnum bonum and Up-to-date which were bred by Sutton. Therefore it became necessary to form special committees on synonymics of potato varieties. An extensive resume on synonymics of potato varieties was compiled by Salaman in 1926.

In order to avoid misunderstandings in the names of potato varieties one should not use unspecific names or terms which may be applied to different varieties with the same shape and color of tubers, such as "Oval white". The name "Stolovyi 19" (table) is also not appropriate because there are hundreds of table-purpose varieties and the number 19, similar to any other number, is easily lost in practice. New varieties must be given clearly worded names without a number before their submission for the State varietal trials.

Appropriately named potato varieties have maintained their names for more than 100 years. Such varieties are Magnum bonum, Early Rose, Burbank and others. Local names of some native varieties from Peru and Bolivia have even greater antiquity, having been retained over a period of more than 300 years, and were related to important cultivated species such as *S. andigenum*, *S. stenotomum* and *S. curtilobum*. On the contrary, unsuccessful or inappropriate names of varieties must be discarded.

During the past few years at the University of Pennsylvania a group of geneticists, under the guidance of P. Grun, conducted a series of studies on the plasma of native cultivated species of potatoes, which are included by us in the subsection *Andinum* (*S. andigenum* and diploid species of *S. stenotomum* and *S. rybinii* which are wrongly termed as *S. phureja*). Moreover, they also studied plasma of the wild species *S. vernei* and *S. sparsipilum* which, according to our opinion, are not the progenitors of the above-mentioned cultivated species.

Regarding the width of the leaf lobes, *S. sparsipilum* is less likely to be a progenitor of diploid cultivated species than *S. multiinterruptum*. The above-mentioned Andean species were compared with the Chilean species from the Pacific coastal region belonging to subsections *Pacificum* and *Chilotanum*. There are clear differences between *Andigenum* and *Chilotanum*. All the improved varieties have been developed on the base of the native *S. chilotanum* and regarding composition of the plasma they are not distinguishable from them in those cases if they are not interspecific hybrids.

Moreover, all improved varieties are morphologically of the type *S. tubero-*

sum which was described by Linnaeus. Therefore, all cultivated varieties should be termed *S. tuberosum* L. But the Chilean native varieties should be considered a geographic form of *S. tuberosum* L., designating them as *S. chilotanum* of the common species *S. tuberosum* L. Hence, *S. tuberosum* L. is represented by secondary, much younger varieties of the Northern Hemisphere and their ancient Chilean initial forms of *S. chilotanum* Hawk.

Plasmas of wild species of the potato in the described geographic zone of the native cultivated varieties, including tetraploid ones, particularly *S. zykinii*, *S. molinae* (*S. guaytecarum*), are distinguishable from the plasmas of *S. chilotanum*. Hence, these wild species are not the direct progenitors of *S. chilotanum*. This means that it is necessary to investigate other possible origins of *S. chilotanum*. One such origin could be the spontaneous polyploidy of the diploid *Mantequilla*, which is associated with *Etuberosa*.

REFERENCES

1. Bukasov, S.M. 1933. *Kartofeli Yuzhnoi Ameriki i Ikh Selektsionnoe Ispol'zovanie* (Potatoes of South America and their use for breeding), Leningrad, pp. 3–146.
2. Bukasov, S.M. 1965. Sistematika dikikh vidov Kartofelya (Taxonomy of wild species of the potato). In the book: *Iskhodnyi Material v Selektsii Kartofelya.* Moscow, pp. 5–23.
3. Bukasov, S.M. 1969. Razvitie ucheniya N.I. Vavilova o sistematike Kul'turnykh rastenii na primere sektsii *Tuberarium* roda *Solanum* (The development of the concept of N.I. Vavilov on systematics of cultivated plants using the example of the section *Tuberarium* of genus *Solanum*). In the book: *N.I. Vavilov i Sel'skokhozyaistvennaya Nauka*, pp. 217–228. Moscow.
4. Vavilov, N.I. 1967. *Zakon Gomologicheskikh Ryadov v Nasledstvennoi Izmenchivosti. Linneevskii Vid Kak Sistema* (The law of homologous rows in hereditary variability. Linnaeus' species as a system). 92 pages. Leningrad.
5. Klyapp, E. 1933. *Kartofel'noe Sortovodstvo* (Potato breeding). Translation from German. 208 pages. Moscow.
6. Kostina L.I. and Z.P. Zholudeva. 1971. Selektsiomye Sorta kartofelya (Improved varieties of the potato). In the book: *Kul'turnaya Flora SSSR.* Kartofel', Vol. 9, pp. 309–384. Leningrad.
7. Rybin, V.A. 1933. Rezul'taty tsitologicheskogo issledovaniya Yuzhnoamerikanskikh kul'turnykh i dikikh kartofelei i ikh znachenie v selektsii (Results of cytological studies on the South American cultivated and wild potatoes and their importance for breeding). *Trudy po prikl. Bot., Gen. i Sel.*, Vol. 2, No. 2, pp. 3–100.
8. Sizova, M.A. 1971. Morfologiya trikhom vidov sektsii *Tuberarium* Bitt.

roda *Solanum* L. (Morphology of the trichomes of the species of section *Tuberarium* Bitt. of genus *Solanum* L.). In the book: Kul'turnaya Flora SSSR. *Kartofel'*., Vol. 9, pp. 385–406. Leningrad.
9. Sinel'nikova, V.N. 1965. Morfologiya lista chiliiskikh form *Solanum tuberosum* (Morphology of the leaf of the Chilean forms of *Solanum tuberosum*). *Trudy po Prikl. Bot., Gen. i Sel.*, Vol. 37, No. 3, pp. 37–49.
10. Yuzepchuk, S.V. and S.M. Bukasov. 1929. K voprosu o proiskhozhdeniya kartofelya (Problems regarding the origin of the potato). *Trudy Vses. S"ezda po Genetike Selektsii i Semenovodstvu*, Vol. 3, pp. 593–611.
11. Yuzepchuk, S.V. 1937. Novye vidy roda *Solanum* iz gruppy *Tuberarium* Dun. (New species of genus *Solanum* from the group *Tuberarium* Dun.). *Izv. AN SSSR, Ser. Biol.*, No. 2, pp. 295–331.
12. Yuzepchuk, S.V. 1962. Problema vida v svete ucheniya Darvina (The problem of species in the light of Darwin's concept). *Sovetskaya Botanika*, No. 6–7.
13. Alefeld. 1866. Landwirtschaftliche Flora von Mittel Europa. 363 p.
14. Bitter, G. 1912–1914. Solana nova vel minus cognita.—Fedde Repert. Spec. Novarum Regni Veg., 10–13. 566 p.
15. Brücher, H. 1960. Problematisches zum Ursprung der Kulturkartoffel aus Chiloe.—*Z. Pflanzenzücht.*, 43, S. pp. 241–265.
16. Canas, A. La Papa. 1901. *Actes de la Societe Scientifique de Chili*, XI, pp. 159–197.
17. Castronovo, A. Las papas chilotas. 1949. Revista de Investigaciones Agricolas, N 3, pp. 209–246.
18. Cevallos, Tovar W. 1914. Noticia historica y classificacion de la papa de Bolivia. *Cochabamba*, pp. 8–97.
19. Correll, D. 1962. The potato and its wild relatives, Section *Tuberarium* of Genus *Solanum*, Texas. p. 606.
20. Dannert, S. 1956. Zur Sistematik von *Solanum tuberosum* L. Die Kulturpflanze. Berlin, pp. 83–129.
21. Dodds, K. 1962. The classification of cultivated potatoes. In: Correll, D. *The Potato and Its Wild Relatives*. Texas, pp. 517–539.
22. Dunall, M.F. 1852. (in De-Candolle) Prodromus (Solanaceae). Paris, 13 (1). pp. 27–387.
23. Gay Claudio. 1849. Historia fisica y politica de Chile. t. V. 76 p.
24. Hawkes, J. 1944. Potato collecting expeditions in Mexico and South America. Cambridge, 142 p.
25. Hawkes, J. and J. Hjerting. 1969. The Potato of Argentina, Brazil, Paraguay and Uruguay. 130 p.
26. Herrera Fortunato. 1931. Variedades de papas cultivades en el Cuzco. *Boletin de la direction de Agricultura Y Ganaderia*, pp. 281–284.
27. Linnaeus, C. 1753. Species plantarum. Stockholm, pp. 184–188.

28. Maldonado, R. 1897. Estudios geograficos hidrograficos sobre Chiloe. Santiago, Ministerio de Marina, 335 p.
29. Molina, J. 1787. History of Chile. Bologna.
30. Montaldo, A. 1962. Las especies de papas silvestres y cultivadas de Chile. *Agriculture Tecnica*, XXXX, pp. 66–152.
31. Ochoa, C. 1955. Expedicion colectora de papas al Norte del Peru. *Biota*, N 34, pp. 1–17.
32. Ochoa, C. 1958. Expedicion colectora de papas cultivadas a la Cuenca del lago Titicaca. Ministerio de Agric., Lima, pp. 1–18.
33. Ochoa, C. 1964. Recuentos cromosomicos y determinacion sistematica de papa nativas cultivadas en el sur del Peru. Lima, 41 p.
34. Ochoa, C. 1965. Determinacion sistematica y recuentos cromosomicos de las papas indigenas cultivadas en el Centro del Peru. Lima, pp. 103–166.
35. Simmonds, N. 1964. Factors in the evolution of *Tuberarium Group. Journal Linn. Soc.*, Vol. 59, p. 376.
36. Soukup, V. 1939. La classificacion de las papas del Departamento de Puno. *Boletin Ministerio de Fomento.* Lima, 18, pp. 3–72.
37. Ugent, D. 1966. Hybrid weed complexes in *Solanum* section *Tuberarium.* Univ. of Wisconsin, 123 p.
38. Vargas, C. 1949. Las papas sudperuanas, Cuzco, I, 144 p.
39. Vargas, C. 1956. Las papas sudperuanas, II, 66 p.
40. Vilmorin. 1902. Catalogue methodique et synonimique des varietes de la pomme de terre. 80 p.
41. Viirsoo, E.V. 1954. Observaciones en papas andinas. *Rev. Agronomica, Noroeste Argentino*, I, pp. 87–98.
42. Walpers, C.G. 1844–1845. Repertorium Botanices Systematicae Lipsiae, t. III, pp. 5–126.
43. Wight, W.F. 1916. Origin, introduction and primitive culture of the potato. Proceedings of the 3rd Annual Meeting of the Potato Association of America. Washington, pp. 35–52.
44. Whitehead, T., T. McIntosh and Findlay. 1953. The potato in health and diseases. Edinburgh, 608 p.

TRUDY PO PRIKLADNOI BOTANIKE, GENETIKE I SELEKTSII

1978 Vol. 62, No. 1

New Wild Potato Species from the Pacific Offshore Islands of South America

V.S. Lekhnovich
Doctor of Agric. Sciences

In 1786, Molina was the first to note the presence of the wild potato in Chile and to mention its local name "malla".

During 1830–1835, the wild potato was discovered on the Chilean offshore islands by English expeditions, including those of Darwin. Its distribution was established up to 50° South. In 1841, Schlechtendahl confirmed the presence of the wild potato along the coast of the Pacific Ocean near Valparaiso. In 1857, Fr. Fronck found the wild potato on Guaiteka Island and Tres Montes peninsula. At the end of the 19th century, Canas reported the existence of wild potatoes in 22 regions on the continent and on the islands off Chile. In 1928, S.V. Yuzepchuk brought to Leningrad two species of wild potatoes in a live state from Chiloe Island; they turned out to be the tetraploids Solanum leptostigma Juz. and S. molina Juz. In 1958, Professor Brücher found a wild triploid S. oceanicum Brücher along the western coast of this island. In 1967, A.G. Zykin collected some wild potatoes along the western coast of Chiloe Island, at 43° South; it was the tetraploid S. zykinii Lechn. Professor Ochoa discovered and collected a nanous growing wild potato-tetraploid S. ochoanum Lechn.—on Guaiteka Island of the Chonos archipelago.

This paper presents a description of the latter two series.

In contemporary science there is no agreement on the wild tuber-forming species of the potato growing in the Pacific, south of the geographic range of the section *Tuberarium* of genus *Solanum*. The major reason for such a state of affairs can be attributed to the lack of a comprehensive study on the wild tuber-forming potatoes of the Pacific coast of South America as well as adjoining islands. A very limited quantity of herbarial specimens have been collected by botanists and an even scantier number of species have been collected and studied in a living condition. There are no references in world literature on the physiographic and climatic history of the southern geographic range of the section *Tuberarium*. There are no results of its study, or lists of the native places of wild tuber-forming potatoes which were recorded by botanists or students of local lore. Moreover, this is the place of origin of our European potato *S. tuberosum* subsp. *europaeum* Buk. et Lechn., and its wild ancestors, which are so needed by plant breeders, grow here. This work is devoted to partial fulfillment of the above-mentioned gaps.

The history of the geographic range. Genus *Solanum* (Tourn.) L. (family Solanaceae A.L. de Jussieu (1789), super order Lamianae (Tubiflorae) Takht., subclass Asteridae (*Agarth*, 1825) Takht., order Scrophulariales Lindley (1833) emerged much before the division of the New World from the Old. Species such as *S. nigrum* L., *S. pseudocapsicum* L., *S. auriculatum* Ait., *S. aculeatissimum* Jacq. grow along both sides of the Atlantic [28]. The genera *Gossypium* L. and *Bromus* L. emerged simultaneously with the genus *Solanum* at the beginning of the Cretaceous period [65]. In geology the beginning of the Cretaceous period relates to the time 137 ± 40 million years ago, which determines the duration of the evolution of genus *Solanum* as not less than 100 million years.

Section *Tuberarium* [(Dun.) Bitt.] Buk. of genus *Solanum* developed in the New World after the emergence of the Atlantic Ocean; its representatives are not found in the wild flora of the Old World. The belief of D. Correll [28] regarding the possibility of the existence of real wild potatoes in the mountains of South Africa at an altitude of about 2000 m is evidently not sound. It is known that isolated European varieties have been cultivated since the beginning of the last century.

Van der Plank [66] was the first to report the presence of wild potatoes in Lesoto and he noticed long stolons on them. Weinmann [69], in his report on the so-called "wild" potato in Basuto, writes that he discovered it in a wild condition at many places in the hilly regions of Lesoto at an altitude of 2000 m among stones and bushes, in between rocks, along river banks, on fallow land and sometimes under cultivation by the local farmers.

Weinmann reports that the potato was introduced into the Cape colony about 1800, most probably from England and Holland. In Lesoto it was spread by the first local missionary, E. Kazalis. This consisted of an assortment of varieties which were grown at that time in Europe. These varieties had colored skin and reddish or yellowish pith in the tubers. In the dispersal of potatoes, Kazalis was greatly helped by a local inhabitant called Raditapole, which means "father of potato".

In 1965 Professor Brücher [25], with the help of workers from the local department of agriculture, collected a considerable number of wild potatoes. He found different types of tubers: round, oval and elongated with red, red-purple, brown, yellow and blue-black skin, with white, whitish-yellow and yellow pith. Chromosome counts on the specimens from Leribe showed their tetraploid nature.

Moreover, Brücher also found potatoes growing wild in Swaziland and Kilimanjaro. He assumed that these were escaped varieties of the previous century from Europe. F. Herold [43], a scientist in the division of virology of the research institute at Caracas (Venezuela), analyzed the specimens which had been collected by Brücher to determine any virus infection. Among 129 plants from the tubers from 29 different places in Lesoto only 8% of them were infected by virus X, 4% by virus Y, 16% by virus S and 3% by the Acuba mosaic. Herold's analysis testifies to the very small occurrence of viruses on potatoes in Europe at the beginning of the previous century and also to the probable existence of virus S at that time.

The comparatively wide contemporary spread of the escaped European varieties of potato can be explained by the seminomadic way of life of the natives during that period, the existence of certain rodents feeding on the tubers and by the participation of birds who dispersed the tubers. Moreover, there were highly favorable climatic and ecological conditions.

Tubers of the wild potato, kindly sent to us from Lesoto by Professor Talukdar, are small (diameter not exceeding 3 cm) and have small eyes. Plants of the winter crop have highly enlarged conical lobes with a slightly dissected (fragmented) leaf and as per type they represent *S. tuberosum* subsp. *europaeum*.

However, in a letter to Dr. Correll [28] Van der Plank reported that among the wild potatoes growing in Basuto there are plants which do not develop tubers and which are, no doubt, much later forms of *S. tuberosum*. Thus confinement of the primary geographic range of wild tuber-forming potatoes to the New World can be considered an established fact.

The emergence of tuber-forming species of Tuberarium. According to the calculations of Core [27], herbaceous plants emerged in the Miocene and developed extensively in the Tertiary period of the Pliocene.

According to Raunkier [1], contemporary tuber-forming species of the section *Tuberarium* belong to a terrestrial cryptophytic type of plant in which not only all aerial parts of the plant, but also some parts of the underground stem die during winter. But the buds sprouting later are situated at a certain depth on the tubers, bulbs or rhizomes. After the Eocene age, i.e. about 60 million years ago, the average temperature of the earth dropped from 21 to

10°C with variations of ± 1°. In the present era the average temperature has again risen by 10° [62]. In the moderate belts of the earth and in hilly regions, the entire plant world, and particularly herbaceous plants, must have developed adaptations for survival during cold or dry periods of the year.

According to Raunkier, other closely related non-tuber-forming species belong to a hemicryptophytic type of plant; one example is *S. lycopersicoides* Dun. from the section *Juglandifolium* Lechn. The evolution of the ancestors of tuber-forming species was linked to the gradual cooling of the environment, as it advanced and spread upwards into the hills where the winter break in vegetative growth is clearly manifested in the tropics. Evolution gradually increased the depth at which the buds formed in the soil apart from the development of the aerial and underground branching of the stem as a result of the absence of apical dominance in the points of aerial growth. Stolons did not begin to bear tubers at once. In *S. chomatophilum* Bitt. from the series *Piurana* Hawk. they now serve as layers (cuttings), growing over the soil surface at a slight distance from the main stem. They hardly ever form tubers in this species, but even if they do develop them, they cannot be stored and they do not play any role in the conservation of plants in their native place (Ochoa, personal report), as well as when grown near Leningrad under long-day conditions. Species *S. chomatophilum* Bitt. is the live evolutionary link between tuber-forming species and non-tuber-forming but already stolon-forming species.

The process of transformation of non-tuber-forming species into tuber-forming ones can be represented in the form of the following mutational situations: 1) increasing the depth of the stem-renewal buds below the soil surface; 2) development of a part of these buds into stolons; 3) emergence of the hormone for tuber formation and development of tuber-like upper thickened portions of stolons, which bear not only the stem-renewal buds, but also reserves of water and nutrients to be used during the primary phases of development of the new plant. Under these conditions two ecotypes developed: one with a dormancy period in the buds of the tubers and one without a dormancy period. The first ecotype is obviously characteristic of tubers of known wild potatoes, while the second is characteristic of tubers of a number of cultivated species grown in hilly tropical valleys. In Chile all forms of the wild potato as well as the cultivated potato have renewal buds in the eyes of the tubers, and have a long dormancy period corresponding to a more severe climate than the contemporary one.

The hypothesis of Professor Hawkes. Hawkes and his colleagues [37] conducted an analysis of the proteins in tubers (not in seeds) of *S. tuberosum* and wild potatoes. A quadrilinear spectrum similar to that of *S. tuberosum* was given by Mexican species of the series *Bulbocastana* Rydb., *Demissa* Buk., *Longipedicellata* Buk. and *Polyadenia* Buk.: *S. acaule* Bitt., *S. bukasovii* Juz., *S. canasense* Hawk., *S. capsicibaccatum* Card., *S. chacoense* Bitt., *S. chaucha*

Juz. et Buk., *S. gourlay* Hawk., *S. infundibuliforme* Phil., *S. lapazense* Hawk., *S. macolae* Buk., *S. megistacrolobum* Bitt., *S. microdontum* Bitt., *S. neohawkesii* Ochoa, *S. neoweberbaueri* Wittm., *S. oplocense* Hawk., *S. raphanifolium* Card. et Hawk., *S. sanctae rosae* Hawk., *S. simplicifolium* Bitt., *S. tarijense* Hawk., *S. toralapanum* Card. and *S. vernei* Bitt. et Wittm. A bilinear spectrum was given by species of the series *Pinnatisecta* Rydb. (*S. jamesii* Torr. and *S. pinnatisectum* Dun.) and *Cardiophylla* Buk. (*S. cardiophyllum* Lindl., *S. sambucinum* Rydb., *S. ehrenbergii* Bitt. Rydb.). A unilinear spectrum was given by the only representative of series *Morelliformia* Hawk., viz. *S. morelliforme* Hawk. The latter is in fact very distant from the rest of the representatives of the section *Tuberarium* Buk. as it is an epiphyte growing in the hollows of trees.

The presence of the representatives of series *Bulbocastana* Rydb. (obviously distant in the system of species of the section as a whole) among species giving a quadrilinear spectrum compels us for the time being to be careful in using the results of immunoelectrophoresis for the proteins in tubers (and not in the seeds) to establish the phylogenetic relations in the section *Tuberarium* Buk.

Unfortunately, some closely related non-tuber-forming species, particularly *S. etuberosum* Lindl., were not included in the analysis. Interpreting the results of immunoelectrophoresis Professor Hawkes [37] put forward a hypothesis on the emergence of section *Tuberarium* of genus *Solanum* in the territory of contemporary Mexico in the late Cretaceous era. Hawkes assumed that before or in the middle of the Eocene a migration of tuberous forms took place from the Mexican primary center to the south in the territory of contemporary South America where numerous secondary centers of diversity developed. After the emergence of the Panama isthmus about four million years ago reverse migration of South American forms took place to the north where their hybridization with the northern forms gave rise to the contemporary series *Demissa* Buk. and *Longipedicellata* Buk.

We must not forget to mention that the species of non-tuber-forming sections (*Muricatum* Lechn., *Caripensum* Lechn., *Appendiculatum* Lechn., *Suaveolens* Lechn.) presently growing in Mexico, are very distant from the series of tuber-bearing potatoes. Much closer to them is the section *Etuberosum* Lechn., which is found only in central Chile, on Chiloe Island, in southwest Argentina, and on the Juan Fernandes Islands.

Only the section *Juglandifolium* Lechn., which is quite distant from *Tuberarium* Buk., has a geographic range from the mountains of Central America (except Mexico) to central Chile. The emergence of non-tuberous species of genus *Solanum* no doubt preceded, and was the substratum for, the development of tuberous species. Therefore, the modern geography of the former species is very important, although it may not coincide with the ancient geography.

As far as mass migrations of plants in the New World is concerned, the flora of the Andes of South America in fact comprises migrant-genera to a great extent. In spite of the complicated geological history of the New World, as a result of the slow development of depressions in the Atlantic Ocean and the process of ascent of the Andean mountain system, boreal and circumpolar flora of the Northern Hemisphere migrated to the Andes of South America, possibly with the help of birds. Many genera were dispersed along the Pacific coasts of North and South America and many migrated from Mexico to Chile and vice versa.

The number of genera characteristic only to the Andes of South America is comparatively limited. According to Weberbauer [67], 34 genera of 20 families comprising boreal flora of the Northern Hemisphere advanced into the Peruvian Andes. Twenty-seven genera of 12 families spread along the Pacific coasts of North and South America. Out of these, ten genera grow equally well in both South and North America, whereas 17 grow mainly in North America. Thirty-five genera of 24 families grow in North or Central as well as South America. In the Peruvian Andes, Weberbauer estimated a total of 47 endemic genera including five genera of orchids.

Harschberger [38], the monographist of North American flora, indicated that the genus *Gilia* includes about 80 species in North America and about 15 in Chile. *Bulnesia*, *Larea*, *Tribulus* and *Drimus* (which extends as far as Cape Horn) are some of them.

Reiche [56] studied the Chilean flora and estimated it to be comprised of 685 genera of Angiosperms from 130 families including 17 genera of the family Solanaceae. About 162 genera are identical to those in California and Mexico, but are represented by different species, which indicates the ancient disruption of links. About 33 families of the flora are found in Chile, but their representatives are not found in California, whereas 22 families are found in California but are absent in Chile.

In this way, the extensive migration of flora from North America to South America and vice versa is firmly established in the modern geography of plants. There is no basis for restricting it only to the narrow and recently emerged Panama isthmus. Carriage by birds was probably the main means of migration of the flora.

The discovery of wild potatoes in Chile. Species of three sections of genus *Solanum* are found in this country and on the off-shore islands: *Tuberarium* (Dun.) Buk. [series *Cuneolata* Hawk. and *Tuberosa* (Rydb.) Buk.], *Juglandifolium* Lechn. and *Etuberosum* Lechn. Species of the last two sections do not develop tubers and their occurrence testifies to the conservation of the tertiary flora here.

Molina [49] was the first to record the occurrence in Chile of the wild tuber-forming potato in his unique monograph on the natural history of Chile. He writes: "In fact, all open places produce them voluntarily in large quantities, but this wild potato, which is called 'malla' by the Indians, is healthy with small bitter tubers.

According to Montaldo [51], the term "malla" in the language of Indians of the Mapuche tribe refers to any wild potato as opposed to the cultivated potato—Ponu or Poni. Brücher's indication of the presence of the word "malla" in the Araukanian dictionary of Andrés Febrés [32] is erroneous, because there is no such word. According to Bayes [51], *S. maglia* has other local names like: Papa zorro, Papa cimarrona and, according to Albert [18], Hierba mora, which when translated means potato "forx", wild potato and dark-lilac grass respectively (the last name was given according to the color of the stems).

In various editions of his monograph, Montaldo mentions [51] that Molina did not give either Italian or Latin descriptions of the wild potato malla (Maglia). An indication by Brücher [24] of the presence of such a description in the German edition is obviously erroneous. Therefore, it is necessary to consider Schlechtendal as the author of the species *S. maglia*. Attribution of authorship of this species to Molina by Gay [35] in his "Flora of Chile" and by Brücher [24] is wrong.

Molina himself was inclined to consider *S. maglia* Schlechtend. as wild *S. tuberosum esculentum* C.B.P., explaining the bitterness of the tubers as due to the flow in cultivation which is of course unbelievable. Simultaneously, Molina [49] mentioned the successful cultivation of 30 varieties of local potato in Chile. He divided the Chilean cultivated potato into two species, common *S. tuberosum esculentum* C.B.P. and *S. cari*, thus indentifying one cultivated form as a separate species and termed it "del suo patria" which might mean some dark forest or the dark color of the tubers (Cari in the Italian language means dark brown and in the language of the Araukans—an important man). He described it as follows: "It has white flowers with large yellow nectar-sacs (anthers—V.L.) inside similar to those in Narcissus and it produces cylindrical tender tubers which are generally fried and eaten". The description in Latin states the following: "Solanum caule inermi herb., foliis pinnatis integ., nectar campanulato subaequante petala" (Solanaceous plant with herbaceous stem, without thorns, with pinnate entire leaves, with bell-shaped nectary, i.e. with anther column almost equal to length of petals).

The assumption by Brücher [24] that the description by Molina of the wild *S. maglia*, under the name *S. cari*, does not correspond to the direct reference by the Italian scientist to the tenderness and edibility of the tubers of *S. cari* (in contrast to the bitter tubers of *S. maglia*) and to the significant length of the anthers—almost equal to the corolla length which is also not observed in *S. maglia*. Probably, the form with white flowers, var. *pachacona* Lechn. (*S. chilotanum* Hawk.) *S. tuberosum* subsp. *chilotanum* Buk. et Lechn., was described. It has horn-shaped or cylindrical tubers, long anthers and filaments with a total height approximately equal to 2/3 of the radius of the corolla;

leaves with narrow lobes, first leaf simple.

In his dictionary Lenz [46] reports that there is a local Chilean variety of potatoes under the name Papas caris which has thin, elongated, index-finger-like tubers with a blue skin and friable sweet pith having anthocyanin veins. It is possible that this particular form was used by Molina while describing *S. cari*.

Pickering [53], the author of the chronological history of plants, also has no doubt that what was described by Molina [49] under the name *S. cari* was the Chilean cultivated potato. In the beginning of his report on the potato, Molina presents briefly information about *S. tuberosum*, relating the Araukan and Mapuche name of the potato—"Poni"—as its synonym. He also mentions the dictionary of natural history by the above-mentioned author and the opinion of Medichi and other authors regarding the Chilean origin of *S. tuberosum*.

S. maglia Schlechtend., the species of wild potatoes which was recorded by Molina for the first time, is still growing despite the increased activity of man. From the data of the monograph of Montaldo [51], the geographic range of *S. maglia* extends in a narrow belt along the coast of the Pacific Ocean up to an altitude of 125 m from Kokimbo (in the north) to the Valparaiso region (in the south).

We find references to the Chilean cultivated potato in the works of Fresier [34], the leader of the French sea expedition to the coasts of Chile and Peru, who found the tubers tasteless (insipid).

The occurrence of the wild potato in Chile on the islands of the Chonos archipelago was also indicated by Darwin [4], who visited the islands in 1834–1835 in the small 260 ton ship "Beagle" and also by Captain Fitzroy [58].

Darwin wrote about the wild potato of Chonos archipelago: "The wild potato grows well on these islands on sandy shell-rocky soil near the sea coast. The tallest among these plants measured 4 feet (1.21 m) in height. Most of the tubers were small, yet I found one tuber of oval shape with a diameter of 2 inches (5 cm); in all respects they were similar to the English potato and even had the smell, but when boiled they became highly wrinkled, watery and tasteless, absolutely without a bitter taste. No doubt, this potato was of local origin. According to Mr. Low, it is found up to latitude 50° South and wild Indians in these places call it Aquinas (the name is probably derived from the Spanish word "aqui"—here). But on Chiloe the Indians call it differently" (4, pages 321–322). Low, referred to by Darwin, worked on the expedition as a pilot on the ship "Adventure" which accompanied the "Beagle" and when the above-mentioned ship went out of order he came over to the "Beagle". Before the expedition, Low was engaged in the seal trade on the islands of South Chile and Argentina, and no doubt he was very familiar with their geography and nature and, therefore, he could discharge the difficult responsibility of pilot in this region. At the time of mapping the islands the northern bay of Guaiteka Island was named after him to honor him by the expedition on the "Beagle".

Obviously, Darwin personally saw the potato on the unnamed island in front of Port Low on Guaiteka Island. This potato was discovered by the captain of the "Beagle", Fitzroy, who did not mention the participation of Darwin in his excursion of January 10, 1835. "Stems, leaves and flowers of this vegetable are of the same size and they seemed equally healthy," wrote Fitzroy in his report, "as those in any English vegetable garden, but the tubers on the roots (stolons—V.L.) were small, watery and tasteless. However, I must remind you that we saw them early in January which corresponds to July (of the Northern Hemisphere—V.L.), many weeks before the time when we may expect edible potatoes in English fields" [58].

The assistant captain of the "Beagle", John Lort Stokes, was preparing maps of the shore of some of the western islands of the Chonos archipelago from December 16 to 31, 1834. Darwin did not participate in this excursion because till December 6 he was examining Chinoe Island and afterward he remained on the "Beagle" till the end of the month studying the coast of the mainland.

In his report on the work done, as cited by Fitzroy, Stokes reported his finding of the wild potato on one of the islands: "On Rowlett Island the potato was found growing wildly; the biggest tuber measured two inches (5 cm) in length and one inch (2.5 cm) in diameter; they were absolutely tasteless [58, pages 2, 374]. Herbarial material for Darwin was obviously taken here.

On January 19, 1830, Lieutenant Skyling from the ship "Adelaide" also recorded the extensive occurrence of wild potatoes to the south of Chiloe Island. On the islands of Guaiyanekas in the region of Spidwell Bay at 47°30′ South latitude he wrote: "In almost every bay we noticed the potato growing amidst wild celery (?—V.L.) close to the level of high flood tide, but displaced by other vegetables it gives very small yield" [58].

Confirming the wild condition of the potato discovered on the islands of Chonos, Darwin sailed through Kordil'era to Mendosa and wrote the following on April 15, 1835, from Val'paraiso to his university Professor Henslow. "Among herbarial specimens from the Chonos Islands you will find a beautiful specimen of wild potatoes growing in a climate absolutely opposite to the climate described just now; no doubt, it is the real wild potato. Certainly it is a species quite different from that growing in the lower Kordil'eras" [29].

Thus, the British expedition composed of the ships "Adelaide", "Adventure" and "Beagle" documented the occurrence of the wild potato on at least three islands of South Chile in the '30s of the previous century and gave reliable information about its occurrence there up to latitude 50° South.

As seen from the photograph of Darwin's herbarium sent to us by Dr. Correll, the upper part of a big plant is preserved in it. Two to three pairs of leaves, without lobes, with very highly increased terminal lobes and with a reduced second pair. The peduncle, with three bostryxes, measures about 10 cm long at the beginning of flowering. Pedicles short, about 15 mm long, with a section of about 5 mm above the joint, with expanded apex, not well separated from calyx. Calyx short. Corolla, obviously big, with radius of about 20 mm.

The picture of this herbarial specimen presented by Correll [28] is not documented but is highly stylized with less sharply increased terminal lobes of the leaves. In the picture, the corolla is distinct with big acute terminal parts, moderately dissected with wide lobes. Unfortunately, numerous later references on the growing of wild potatoes in Chile and its islands, as presented below by us, are not confirmed by herbarial specimens of the latter, but often they are accompanied by a reference to the use of tubers of wild potatoes as food for man, which is extremely significant; that is why similar references can also be considered as thoroughly reliable [6–8].

It follows from the above that the islands and coasts of Chile are of great interest for the study of species of the series *Tuberosa* Buk. [3]. They have not been studied fully. This lacuna must be removed as soon as possible. From the above-mentioned quotation of Darwin it is evident that the Indians of Chiloe Island knew about the existence of wild potato and even had a special name for it. It is doubtful that the source of their acquaintance with the wild potato was the wild growth of the latter on the islands of the Chonos archipelago which are separated from Chiloe Island by about 50 km of open, and in these places turbulent, ocean. No doubt, what was referred to by the Indians was the local potato of Chiloe Island. The discovery of the wild tuber-forming potato on the islands of South Chile by the British expeditions during 1829–1834, as well as the finding of this wild potato by Fr. Fronck, remained unknown to the Chilean monographer of potato, Canas Pinochet Alexandro [26], who during his travels through the country collected reliable information about the growth of the wild tuber-bearing potato in the provinces of Concepsion, Arauko, Mal'eko, Kautin, Val'diviya and L'yankunue and on Chiloe Island along with a few coastal small islands.

Although the list of the locations of sightings of the wild potato compiled by Canas [26] is not supported by herbarial specimens, it should not be considered unreliable because the sources of information for the compilation of the list were reliable local natives, and soldiers who had personally seen wild tuber-bearing potato plants and who had been using their tubers for food. We shall present the complete list of Canas:

Place of growth of the tuber-bearing potato	*Source of information*
On the hills of Nahuelbuta and between the rivers Bio and Imperial.	Information obtained from the writer Benkhanin Vikuna Matskene.

Last Cayucupil at 10–15 km in the southwest of the city Canete.	Personal excursion of Canas and botanist Rikardo Galindo (3.10.1895).
Arauco Province (a species with small tubers which are not used for food grows here).	Report of the former governor of this province, Frederic Kilo.
On the island and banks of the mighty river Cautin of the same province near the town Temuco. Grows in abundance.	Soldiers of Colonel Alekho Karilo, 1882.
On the hills near the fort 40 km south of the town Temuco.	Detachment of Colonel A. Karilo, 1883.
On the Dona Ines Island in the Imperial River, 54 km from its estuary. Grows in plenty and is used for food by local Indians in famine years. Note: the Imperial River begins after confluence of two rivers, viz. Cautin and Cholchol.	Work of captain of the corvette, Leonsio Sinoret: "Reconocimiente del Rio Imperial. Annales de la Universidad. Julio. 1870" (Research on the Imperial River).
On the coastal hills and among bushes, on the slopes of the Cordillera in the provinces of Valdivia, Arauco, Cautin.	Information concerning Valdivia Province from Colonel Alekho Karilo.
Around the mountains Corral, Cordillera of Chanchan, near the town Tolten in Cautin Province.	Indians call "Cachito" due to curved shape and tender taste of the tubers. There are toltena blanca and colorada, i.e. white and colored tubers.
Among bushes on the slopes of the Cordillera in the hollow of Lake Ranco, between lakes Panquipulli, Pirihuaico and Lacar in Valdivia Province.	
On the slope beside Villarica and Pucon. In the hilly approach to Maichi Cogurga.	
Around the town Calbuco, on the northern coast of Ankund Bay. Department Carelmapu of Llanquihue Province.	Information from Manuel Andrade, the governor for life in Calbuco. Potato is called Guapa. Tubers have red veins.
At Lake Llanquihue of the same province.	Rosales Visente Peres in the work "Ensayo sobre Chile Reino vejetal", 1852.

Chiloe Island	Information from Pedro Jose Andrade, the municipal representative in Chonchi.
Tablacura, 40 km south of the village Cucao, western bank of Argentina, on the southeastern shore of the island not far from the split of San Pedro Island extending to the south. In the village Antao, in the estuary and along the banks of the river of the same name flowing towards the west.	
At many places along the same banks and at Cape Quilan. On Chahualin Island opposite the port of Queilen. On Lobos Island, in the north of Chiloe Island.	Information from Nicolas and Paulino Alvares and Felix Nieto, treasures of the town Castro.
On Chulin Island from the Desertos group of islands in the same department.	Information from Juan F. Carrer from the island Kal-macha, a rejidor of the municipality of Kenes.
On rocky island Pedra de Caltos from the group of islands in Quinchao department.	
Numerous natural plant populations on the Huaitecas islands, particularly on the islands Huamblia and Cabras under the name "Oka".	Information from the cultivators of Karelmapu.

Besides Canas, Chilean scientists working on the flora also recorded the growing of wild tuber-bearing potatoes, mostly on the mainland.

Gay [36], the author of the "Flora of Chile", recorded the presence of the wild potato *S. tuberosum* L. in Araucania on the ridge of Malvarco, on the "potato" ridge (Poñis), and on the hillocks around Santiago, according to the report of Valdivii, the conqueror of Chile.

Reiche [56, 57], a florist of Chile, mentioned the spread of wild *S. tuberosum* L. on the lower ridge of the Cordillera in Aconcagua Province up to Llanguihue, on the Linares Paral ridge where we have a number of "potato" mountains (Poñis), near the lagoon of Malvarco, in Argentina, on the slopes of the Cordillera in the provinces of Rancagua and Colchagua in central Chile.

It seems strange but not a single European expedition seeking initial material for breeding potato systematically examined the regions indicated by Canas and other florists, where in the past century there were bushes of wild potato tubers.

Chiloe Island[1]. Here lies the genetic center of diversity of the native cultivated potato (*S. chilotanum* Hawkes) *S. tuberosum* subsp. *chilotanum* Buk. et Lechn., discovered by the Soviet expedition under the guidance of S.V. Yuzepchuk. This island is the main native home of the European potato *S. tuberosum* subsp. *europaeum* Buk. et Lechn. Therefore, the results of the study of this island, its geology, history, flora, geography, climate, flora of wild potatoes, and ethnography are of great interest.

Darwin was the first scientist of wide background and varied interests to visit this island. Unfortunately, he did not notice its wild populations of potato, which no doubt were more extensively found than at present. The reasons for this were the selection of inappropriate routes and an unsuitable period of time for the visit to that island, as may be seen from the following account of his journeys. Apart from these, the most important reason was the lack of sound knowledge of the Spanish language which could be understood by the local Indians.

Darwin visited Chiloe Island thrice in 1834 and in the beginning of 1835. The abundance of cultivated potatoes did not escape his attention, even during the first brief visit around July 10, 1834, at the time of the greatest severity of winter in the Southern Hemisphere, where the "Beagle" remained anchored for many days, obviously in the bay at the village Cucao. On July 20, 1834, Darwin wrote to his sister about this visit to Chiloe Island: "We left Chiloe Island a week ago; a number of succeeding storms compelled us to visit this place. We stayed there for a few days so as to strengthen our people. Potato there was in plenty as in Ireland. But for the exception of this weighty advantage, Chiloe is an unfortunate hole with regard to its climate (30, page 100).

The second visit to Chiloe Island continued from November 21 to December 10, 1834, when seedlings of the wild potatoes had already emerged. Darwin did not notice them either on Chiloe Island, or on Lemui Island.

The third time Darwin went to Chiloe Island was from January 22 to February 4, 1835, when seedlings of wild potatoes had already developed. Wild potatoes again escaped his attention.

Darwin wrote the following to his sister Catherine on March 10, 1835, about the study of Chiloe Island: I examined Chiloe thoroughly going around on horseback in both directions (30, pages 100, 110, 114).

During Darwin's time the only inhabited place on the western coast of Chiloe Island was Cucao, where over a stretch of 4–5 miles (6–7.5 km) along the coast lived 30–40 Indian families. The eastern coast of Chiloe was somewhat more densely populated.

Darwin was told on Chiloe Island that an omnivorous bird *Polyborus chi-*

[1]According to Fitzroy, the captain of the "Beagle", the name of the island, Chiloe, is derived from two Arukanian words: Chile-hue "End of Chile" [58].

mango can pull out and eat (and this means disperse) the newly planted potato tubers causing serious damage to the crops [4]. Darwin had been on the Tres-Montes peninsula in the south of Chiloe on December 30, 1834, but obviously because of the short period of stay at the beginning of the summer in the Southern Hemisphere he did not notice the wild potato which was collected by Fonck later on in March 1857. Correll wrongly related the mainland peninsula of Tres-Montes to the Chonos archipelago [28]. But, no doubt, it is the same floristic province.

The existence of a bird which can pull out the newly planted potato tubers, as established by Darwin, is of great interest, because it can explain some migrations of species of the wild potato.

The climate of Chiloe and its more southern islands is oceanic. The west coast of Chiloe Island is characterized by an average maximum temperature of 16–17°C in the hottest month and the lowest temperature is +11°C. The average minimum temperature during summer months is 10–11°C and during winter months +5° [45]. Frosts occur along the entire coast and the farther inland we move, the more severe they become. Rainfall occurs mostly during winter, whereas the total amount of rain in the summer months is 60–70 mm.

Montaldo [51] gives the following data regarding the average annual rainfall and the average annual temperature for different places on Chiloe Island and on other islands situated to the south of it:

Geographic point	*Amount of rainfall* (mm)	*Temperature* (°C)
Chiloe Island		
Ankud	2384	10.7
Kutal'makhue	2500	11
Puntra	3000	11
Puiukhuapi	3810	10
Dal'kakhue	2200	11
Pirukina	2530	11
Castro	2070	11
Yuzhui	2000	11
Cucao	2500	10
Guafo Island	1391	10.2
Guaitekas Island of Chonos Archipelago	4277	9.9
Guaiteka Island, Port Lou	3000	10.1
Chonos Archipelago	4500	9.4

It is evident from the above-mentioned values for the annual rainfall and temperature that both wild and cultivated potatoes on Chiloe Island and its southern islands are ecologically plants of a cool and humid climate, but the contemporary era has a much milder climate than the Pleistocene period.

Herbarial specimens of wild tuber-bearing Chilean potatoes. As stated above, Molina [49] in his monograph mentioned the growing of triploid *S. maglia* Schlechtend. in Chile, but he did not describe this species in any other editions of the book. In 1928 S.V. Yuzepchuk found two species of the wild tuber-forming potato—*S. leptostigma* Juz. and *S. molinae* Juz.—on Chiloe Island and brought them to VIR (All-Union Institute of Plant Industry). These species were replanted in his vegetable garden by a local inhabitant named Crysti.

During the last 20 years three more species have been discovered. One of them (*S. oceanicum* Brüch.) was described by Brücher [24]. Two other species discovered by Zykin [6, 7, 64] on Chiloe Island and by the Peruvian scientist Ochoa on Guaiteka Island are described below.

S. zykinii Lechn. sp. nov.

Plantae stolonibus longissimis usque ad 90 cm, die longo pro parte, caules novos formantibus, tuberibus rapiformibus vel oblique rapiformibus, roseis, circa 35 × 50 × 40 mm, carne alba, amarella.

Tuberum blastemates basi rotunda, rubra, pilis erectis obsita, apice semieffuso, pilis densis prostratis instructo. Caules brunei alis subrectis instructi in insula Chiloe solum circa 50 cm, in Lenopole circa 60–70 cm altae, internodiis circa 15 mm longis, foliis 3–4–5 jugis, margine undulato, foliolis interjectis paucis 1–3, vel in Chiloe nullis, foliolo terminale majore circa 64–113 × 66–65 mm petiolulo 14–17 mm longo, basi cuneatulo, apice elongato, foliolis jugorum circa 27–84 × 16.5–47 mm, petiolulo circa 4 mm longo, basi, inaequali, foliolis interjectis parvis ca 8 × 7 mm, petiolulis brevibus, alatis.

Pedunculi axillis 14–15 crescentes, 130–200 mm longi pilis brevibus sparsis prostratis obsiti, furca elongata, inflorescentiis latis, multifloribus—(12–27 floribus), pedicellis 22–26 [30] mm longis, articulatione 8–10 mm infra calycem, apice valide incrassatis, pilis sparsis, brevibus prostratis instructis.

Calyx parvus, pedicelli apicis debiliter limitatus dissectusque, sepalibus ca 10–4 × 3–8(5) mm, basi valide glumaceis, mucronibus subularibus, alabastro patentibus, pilis sparsis prostratis obsitus. Alabastrum conicum, corollae acuminibus breviter pilosis. Corolla alba extus stellae radios longitudinaliter subluteola, debiliter dissecta, petalis (laciniis) ca 16–18 × 22–18 mm, lobo brevi, solum ca 2 mm lg, acumine mango ca 5–6 × 6 mm, nonnunaquam plena, petalorum commissura undulata.

Filamenta brevia, ca 2 mm longa, lata ca 1.5 mm Antherae latae, basi debiliter curvatae sinuataeque trapecioideae ca 6–7.5 × 2.5–3 × 1.5–2 mm, sulco dorsali angusto, polline numeroso, columna antherarum compacta ovalis.

Stylus ca 12–13 mm longus, valide ad 5 mm antheras superans, rectus, ca 0.5–0.6 mm crassus, basi papallis instructus. Stigma claviformis, profunde sinuatus, non profunde sulcatus, stigmate *S. leptostigma* obtusior.

Ovarium oviforme. Baccae virides, breviter ovales, debiliter compressae, ca 15×14×15 mm, quanto numerosae.

Typus: Dr. A.G. Zykin: WIR k-7600 in statu silvestri insula Chiloe, 43° oceani litore, Ochoa, Ancud, sepulcreto k-11288.

Affinitas. A.S. oceanicum Brücher: caule valide humiliore, pedicelli articulatione valide altiore, corollis, antherisque minoribus, chromosomatum numero (S. oceanicum 2n=36). Similia sunt: corollis debiliter dissectis, acuminibus magnis. A.S. leptostigma: Tuberibus rapiformibus, in statu sano non elongatis, caule crassiore internodiis brevioribus, foliorum petiolis longioribus, foliis minus dissectis, foliolisque majoribus, latioribus, foliolorum marginibus undulatis, pedunculis pedicellisque longioribus, articulatione altiore, calycis sepalis longioribus, corollae petalis majoribus, antheris longioribus latioribusque, stigmate latiore, fertilitate majore. Similia sunt foliorum apicibus elongatis, pedunculi pedicellique pilis sparsis, brevibus, prostratis filamentis brevibus latisque, stylis longis, stigmatibus claviformibus.

Stolons very long, in healthy plants up to 90 cm long, pinkish, under longday conditions near Leningrad they partly come up to the soil surface and produce shoots with rich foliage. Tubers rapiform or oblique rapiform, pink, clearly pink towards the onset of spring, under cultivation, measuring up to 35×50×40 mm, whereas the original tubers measure 15–18 mm in diameter with white watery pith. They have an unpleasant taste of solanine (Fig. 1, a and b).

The aerial shoots with a round red base, wide neck and green apex, tall pink root nodes, semieffusive apex of key-shaped constricted leaves. Pubescence of the base erect, thin, rare, but that of the neck and apical leaves—dense, spreading. Plants with average, vigor, at their original native place about 50 cm, near Leningrad in open field—up to 60–70 cm tall, with profuse foliage, internodes in the middle part of the plant about 15 mm long, with numerous auxiliary shoots. Stem brown, winged, with almost straight wings of about 3–4–5 mm height. Leaves (upper and middle) attached to stem at an acute angle, while the lower leaves are horizontal, not dense, and measure about 127–302 ×78–60 mm, with long thick flat petioles measuring 57–40 mm, generally in four pairs, rarely three pairs, without lobes. Margins of the young lobes of leaf prominently undulated. Terminal lobe round or round-oval, bigger than lateral lobes, particularly on young leaves, measures about 64–113×66–65 mm, on a stalk measuring about 14–17 mm long, with slightly wedge-shaped base. Lateral lobes also big, oval, with elongated acute apices of dropping type, often with unequal bases, on short stalks of about 4 mm length and measure about 27–84×16.5–47 mm. The first pair of lobes is sometimes smaller than the second pair and it is inclined towards the terminal lobes with an acute angle. The differences in the size of the terminal and lateral lobes narrow down in much older leaves. Leaf lobes stiff, hard in touch. Leaf petiole covered with big hairs from the lower side. Lobes small, round, measuring about 8×7 mm on short winged stalks measuring about 1 mm. Peduncle

Fig. 1. *S. zykinii* Lechn.: a—the general appearance of the plant; b—parts of the flower.

develops from the axis of the 14th–15th leaf, brown, with a length up to 130–200 mm towards end of the flowering, pubescent with rare short bent hairs. Branches of peduncle grow intensively at the time of flowering; this is why the old inflorescence is not compact, with displaced peduncles of involucres. Involucres 2–3 having 6–7 (9) flowers in each. Pedicles short, about 22–26 (30) mm long, with the portion above the joint 8–10 mm, of average thickness, highly thickened towards the apex, rarely pubescent with bent short creeping hairs. Calyx small, slightly dissected, with highly glumaceous bases of the lobes, not distinctly separated from the apex of pedicle, with lobes about 10 mm long, bases measuring about 4 × 3 mm and subulate mucrones remaining in the bud, length about 3–5 mm. Pubescence very short, spreading, rare. Bud conical, with poor protogyny of style and stigma, short pubescence of the acute lobes of corolla from outside and bright cream-colored lobes of corolla around stellar rays.

Corolla white, in the bud creamish on the outer side, particularly along stellar rays, very slightly dissected, with lobes measuring 16–18 (22) × 22–18 mm, with oblique short lobe measuring only up to 2 mm in length and with a big acute terminal lobe measuring about 6 × 6 mm. Double petalled corolla is also found (evidently under developed secondary corolla in dry summer). The joining line of the lobes of corolla is goffered. Filaments short, wide, width about 1.5 mm.

Anthers in a compact oval column, slightly curved, wide, trapezoidal, with a small intersac depression at the base of the back, with narrow clear ridge on the back, length 6–7.5 mm, width at the base 2.5–3 mm and at the top 1.5–2 mm. Pollen abundant. Style very long—length 12–13 mm, thickness 0.6–0.7 mm, highly projecting—up to 5 mm above anthers, covered with papillae in the lower part. Stigma club-shaped, with shallow ridge, deep sinus at the base but it is somewhat more obtuse than in *S. leptostigma* Juz. Lobe of the stigma oval from the side. Ovary ovate, sharp. Flowers begin to close at about 15 hours according to solar time. Berries green, short-oval, slightly oblate measuring about 15 × 14 × 15 mm. When seeds are sown we do not notice any significant segregation. Collected by Dr. A.G. Zykin on the western coast of Chiloe Island, at about 43° South latitude, 20–25 km south of a small fishermen's village, Cucao, on the bank of a small river at 80–100 m from the shore of the ocean on pebbly, sandy soil, in the absolutely unhabitated region of Chiloe Island.

Type: Specimen WIR k-7600. An identical specimen, but infected by virus, was discovered by A.G. Zykin among fertilized Indian varieties of potato in the region of a deserted hamlet in Cucao village. One more specimen was found by Professor Ochoa in the burial ground of the town Ankud on Chiloe Island (k–11288). Highly susceptible to *Phytophthora*. According to preliminary data, it is resistant to the Dalem race of potato wart.

Affinity: Clearly distinguished from *S. oceanicum* Brüch., by much shorter (approximately twice) stem, absence of any splitting off of the terminal lobe of the leaf (under cultivation), poor undulated nature of the wings of stems, presence of anthocyanin coloration of tubers and stem, much higher position of articulation (joint) on the pedicle (in *S. oceanicum* at a height 1/5 the length of the pedicle or lower), much longer mucrones of corolla (*S. oceanicum* 1–2 mm), smaller size of less dissected corolla not measuring even 35–40 mm in diameter, much shorter anthers not measuring 8 mm as in *S. oceanicum*, more frequent berry formation, habitat 25–28 km further north, but also along the shore of the ocean, chromosome number $2n=48$, while in *S. oceanicum* $2n=36$ according to Brücher.

On the basis of the photographs of the herbarial material sent to us by Dr. Correll, they are similar to each other in regard to poor dissection of the corolla and large acute terminal lobes.

It is distinguished from *S. leptostigma* Juz. by the following characters: transversely oval not elongated tubers, under healthy conditions having anthocyanin coloration, much thicker, about 15 mm stem, shorter internodes, longer (than 2–3 cm) petiole of the leaf, less dissected, often somewhat larger leaves, undulated edges of bigger and wider lobes, shorter stalk of the terminal lobe, longer peduncle and pedicels, somewhat higher position of articulation (higher than half), longer lobes of calyx, bigger lobes of corolla, longer and wider anthers, wider stigma, more frequent berry formation. These are similar in respect of short and wide filaments, very long styles, club-shaped stigma, creeping rare pubescence of peduncle and pedicel, elongated apex of the lobe of the leaf. This points out their ancient ecological adaptability to the climate and to excessive rainfall.

S. ochoanum Lechn. sp. nov.

Plantae stolonibus, pro parte die longo ascendentibus, caules novos formantibus, tuberibus planiovalibus, cyaneis violaceis ca $35\times30\times10$ mm.

Tuberum blastemates cyaneiviolacea, rotunda, pilis erectis tenuibus obsita, apice semieffuso, pilis densis, prostratis instructo.

Caules conditione edaphica sicca solum 20–30 (40) cm longae conditione humida ad 60 cm altae, basi ad 20 mm crassae non undulate alatae, foliis ca $145–175\times100–130$ mm, petiolis 30–45 mm longis. Folia 4–6 juga, 1–16 foliolis interjectis, foliolo terminali vix majore ca $48–60\times31–37.5$ mm, petiolulo 18–20 mm longo, foliolis jugorum geminorum oviformibus ca $55–65\times32–35$ mm, basi basiscopico inaequalibus, petiolulis 2–10 mm longis, apicibus elongatis.

Pedunculus 60–140 mm longus, crassus (ad 4 mm), 2–4 circinnis, 7-11 floribus circinno.

Alabastrum cylindricum. Pedicelli ca 2 mm crassi, apice valide ad 2.5 mm incrassati, 18–25 mm longi, articulatione ca 5–7 mm infra calycem, saepe anthocyano pigmentata, pilis brevibus $\pm$densis, prostratis. Calyx parvus, saepe sexisepalis, 8–10 mm longus, debiliter dissectus basibus ca 3×3 mm, mucronibus subularibus ca 2–4 mm longis, alabastro recurvatis, pilis semiprostratis instructus.

Corolla saepe sexipetalis, petalis ca $14–21\times14–20$ mm, valide dissecta cyanoviolacea vel alba, lobis elongatis usque ad 8 mm longis, acuminibus parvis ca 3×3 mm, nonnunquam 5×6 mm.

Filamenta ca 4 mm longa, angusta ca 1 et minus mm lata. Antherae ca $7–8\times2.5–3\times1.5$ mm, late dorsaliter sulcatae, columna cylindrica vel debiliter conica. Stylus 9–13 mm, late longus, usque ad 3 mm antheras superans.

Stigma compressicapitatus, saepe profunde sulcatus. Ovarium obtusum, intus saepe anthocyano pigmentatum. Baccae rapiformes ca 16×13×13 mm.

Typus: Professor Carlos Ochoa. Chile, Insula Guaiteca. WIR k-11289, k-11290.

Affinitas. A.S. Zykinii: Anthocyano cyaneo, tuberibus elongatis, conditione sicco plantis minoribus, foliis corollisque dissectioribus, filamentis longioribus, angustioribus, stigmatis baccarumque forma.

A.S. molinae Juz: plantis minoribus, foliis dissectioribus pedicellis longioribus, pedicelli articulatione altiore, corollis minoribus.

A.S. tuberosum var. guayetecarum (Bitt.) Hawkes, Fonck N 102 C [37]: folio dissectiore, minus compacto, corollae petalis dissectioribus latioribusque, filamentis angustioribus.

A.S. specie Cl. Darwino insulis Chonos collecto (ut videtur insula Roulett per Stokes praesentato): plantis valide minoribus, foliis dissectioribus, corollae accuminibus, minoribus, formaque petalorum.

Stolons long, up to 50–80 cm, under long-day conditions of Leningrad some of them come up to the soil surface and form late shoots with profuse foliage (Fig. 2, a and b).

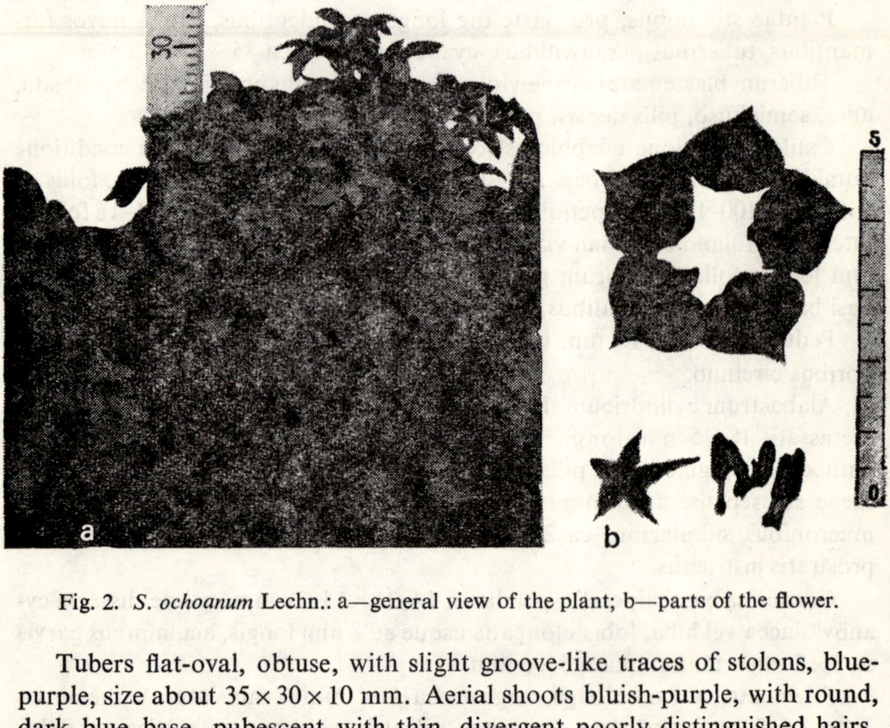

Fig. 2. *S. ochoanum* Lechn.: a—general view of the plant; b—parts of the flower.

Tubers flat-oval, obtuse, with slight groove-like traces of stolons, blue-purple, size about 35×30×10 mm. Aerial shoots bluish-purple, with round, dark blue base, pubescent with thin, divergent poorly distinguished hairs, with a long internode and slightly involuted (weakly pigmented) apical part comprising carinately bent leaves which are covered with dense creeping hairs. Root nodes quite numerous, blue-purple. Under drought conditions plants are very low statured—20–30 (40) cm, but in humid summer—up to 60–70 cm tall, exceptionally densely foliated, with semi-prostrate, very thick (about 20 mm in diameter) main ungoffered winged stem, with very short internodes, with quite numerous lateral shoots developing from all over the axis. Leaf axis and partly the stem also generally colored with anthocyanin. Leaf with long petiole, length about 3–4, or 5 cm, and measures about 145–175 × 100–130 mm with the petiole. Dissection of leaf varies from 4 to 6 pairs of lobes and from one to sixteen pairs of lobelets all displaced to the stalks. A more dissected leaf has the formula:

$$5\text{–}6\ K_{2+1}\, I_{3(\bar{\perp})}\, II_{2\,(\bar{\perp})} + _{(\bar{\perp})} \text{ less than or } {}_{T}III_{2\,(\bar{\perp})} + _{(\bar{\perp})} + 1 < ^{IV}{}_{0} \text{ or } 1 < .$$

In other words the leaf has 5–6 pairs of lobes: in the final series—two pairs of smaller lobes and one pair of lobelets; in the first series—three pairs of smaller lobes, of which one pair is basiscopic and the second is acroscopic, on the stalk of the lobes; in the second series one pair of acroscopic and one pair of basiscopic smaller lobes plus the same number of lobelets and one more lobelet displaced in the axis of the stalk or only one pair of acroscopic lobelets; in the third series, two pairs of small lobes including one acroscopic pair, displaced on the stalk of the lobe, plus one pair of lobelets, acroscopically displaced on the stalk, plus one pair of lobes of the fourth order displaced in the axis of the stalk; in the fourth series smaller lobes are absent or only consist of one pair displaced in the axis of the stalk.

The terminal lobe is just a little bigger than the lateral lobes, measuring about 48–60 × 31–37.5 mm on a petiole measuring 18–20 mm in length. The lateral lobe is ovate, measuring about 55–65 × 32–35 mm, basiscopically not equally lateral, on 2–10 mm long stalks. Lobes of young leaves have elongated apex. Lobelets small on stalks. Peduncle up to 60–140 mm long, and projects considerably above the low-statured plant, very thick (up to 4 mm in thickness), with 2–3–4 bostryx (whorls) with 7–11 flowers in each. Bud cylindrical, often with considerable protogyny. Pedicel thick, up to 2 mm in diameter, highly (up to 2.5 mm) expanded at the apex, not very clearly or not at all separated from calyx, length 18–25 mm, with a section above articulation about 5–7 mm, generally with anthocyanin colored articulation or joint entirely or only below articulation pigmented with anthocyanin, pubescent with more or less dense, bent, short, creeping hairs.

Calyx often with six lobes, slightly dissected, small, with lobes measuring 8–10 mm long and having a base of about 3 × 3 mm and subulate mucrones, 2–4 mm long, curved in the bud. Covered with semi-pressed hairs. Corolla often with six lobes, rarely double petalled, very highly dissected, with lobes measuring about 14–21 × 14–20 mm, with a long lobe measuring up to 8 mm in length, a length of the line of fusion 8–12 mm and small acute ends 3 × 3 mm, but we may find much bigger acute ends measuring up to 5 × 6 mm. Stellar rays green. Filaments up to 4 mm long, narrow, width about 1 mm or less. Anthers in a slightly conical column of about 7–8 × 2.5–3 × 1.5 mm with a wide shallow ridge on the back. Style 9–13 mm long, of average thickness, projects above anthers up to 3 mm. Stigma acute, capitate, often very deeply ridged and not unfrequently trilobed. Ovary obtuse, often colored with anthocyanin in transverse section. Berries roundish, initially rapiformed, size 16 × 13 × 14 mm.

When cultivated it gives segregation in respect of the degree of dissection and color of the leaf, color of corolla (white flowered) and stem, maintaining the small size of the plants, big dissection of corolla, narrow filaments and other characters.

Type: Collected by Professor Ochoa (Lima) on Guaiteca Island of Chonos archipelago. WIR k-11289 and k-11290. Highly susceptible to *Phytophthora*. According to preliminary trials, specimen k-11290 is resistant to the Dalem race of potato wart.

Affinity: It is distinguished from *S. zykinii* by blue anthocyanin, oblong tubers, short-statured plants, more dissection of leaves, presence of splits of the terminal lobe, considerably more dissection of the corolla, with much wider lobes, narrower and longer filaments, much lower, sometimes very deeply bifid stigma, much smaller berries.

From *S. molinae* Juz. it is distinguished by smaller size of better foliated plants, more dissection of leaves, longer peduncles, much higher position of articulation, smaller size of corolla and a less conical antheral sac.

From *S. tuberosum* var. *guaytecarum* (Bitt.) Hawk. (Fonck N 102 C) [40] it is distinguished by less dense, often more dissected leaf, much wider lobes of a more dissected corolla, absence of projecting pubescence of calyx, much narrower filaments, and does not have a club-shaped stigma.

From the herbarial specimen that Darwin collected in December 1834, on one of the islands of the Chonos archipelago (probably, on Roulett Island through officer Stokes), and related by Professor Hawkes to the above-named botanic variety of *S. tuberosum*, judging from the picture presented by Correll, the comments of Darwin and the photograph of the herbarium, it is distinguished by the aspects of small-statured plants, i.e. by shorter stem, more dissected leaf, marginally increased terminal lobes of the leaf, absence of projecting pubescence of calyx, which was shown in the picture by Correll, much smaller size of acute conical corolla, and a compact but not cracked anther sac as in *S. maglia*.

From *S. tuberosum* var. *chubutense* (Bitt.) Hawk. [40] judging from the picture presented by Correll, it is distinguished by flat-oval tubers of smaller size, wider lobes of leaves and absence of wings on the stalk, absence of projecting pubescence of calyx, shorter style, less projecting stigma above anthers, a stigma with a ridge, and a conical anther sac. They are similar with respect to the average wide filaments and dissection of leaf.

From *S. maglia* Schlechtend. it is distinguished by the absence of a highly increased terminal lobe of leaves, marginal pubescence of the lobes of leaves, occurrence of the tissue of the anther on the filament of the burst anther sac. They are similar with respect to the deep ridge on the style.

Further studies of the islands of South Chile, as well as of the mainland, will no doubt reveal a number of new forms of wild potatoes. It is necessary to expedite this work because the growth of population, expansion of cultivable land and increase in the number of cattle fatally affect many representatives of wild flora.

REFERENCES

1. Alekhin, V.V. 1944. *Geografiya Rastenii* (Geography of Plants), pp. 62–63. Moscow.
2. Bryukher, G. 1968. Tsitogeneticheskoe i morfologicheskoe sravnenie klubnenosnykh vidov *Solanum* S beregov Tikhogo okeana i Andiiskikh gor (Cytogenetic and morphological comparison of tuber-bearing species of *Solanum* from the shores of the Pacific Ocean and the Andeas mountains). *Genetika*, No. 2, pp. 80–89.
3. Bukasov, S.M. 1971. Dikie vidy kartofelya (Wild species of the potato). In the book: *Kul'turnaya Flora SSSR*, Vol. 9, pp. 5–40. Leningrad.
4. Darwin, Charles. 1953. *Puteshestvie Naturalista Vokrug Sveta* (Journey of a naturalist around the world), pp. 320–321, 546–547. Moscow. Also see: *Poln. Sobr. Soch.* 1953, Vol. 1, pp. 253–331; 1959, Vol. 9, p. 86. Moscow.
5. Zhukovskii, P.M. 1959. Po tsentram proiskhozhdeniya kul'turnykh rastenii Latinskoi Ameriki (Through the centers of origin of cultivated plants of Latin America). *Botan. Zhurnal*, No. 2, pp. 732–737.
6. Zykin, A.G. 1971. O dikorastushchikh sorodichakh chiliiskogo Kartofelya (On the wild relatives of the Chilean potato). In the book: *Rastitel'nye Resursy*, Vol. VII, pp. 242–251. Leningrad.
7. Zykin, A.G. 1973. Chiliiskii kul'turnyi kartofel' i ego dikie sorodichi (The Chilean cultivated potato and its wild relatives). *Trudy po prikl. Bot., Gen., i Sel.*, Vol. 50, No. 3, pp. 216–240.
8. Lekhnovich, V.S. 1971. *Solanum tuberosum* L. subspecies *chilotanum* Buk. et Lechn. In the book: *Kul'turnaya Flora SSSR*, Vol. 9, pp. 225–304. Leningrad.
9. Markov, K.K. 1951. *Paleogeografia* (Paleogeografia). Moscow.
10. Monin, A.S. 1977. Istoria zemli (History of the Earth). p. 142. Leningrad.
11. Nemkov, G.I., M.V. Muratov, I.A. Grechishnikova, V.A. Gustomesov and others. 1974. *Istoricheskaya Geologiya* (Historical Geology). Moscow.

12. Strakhov, N.M. 1960. *Istoricheskaya Geologiya* (Historical Geology). Moscow.
13. Takhtadzhyan, A.L. 1966. *Sistema i Filogeniya Tsvekovykh Rastenii* (System and phylogeny of flowering plants). Moscow-Leningrad.
14. Ushakov, S.A. 1976. *Geofizika o Dreife Materikov* (Geophysics of the Drifting of Continents). p. 16. Moscow.
15. Ferkhtugen, Dzh., F. Terner, L. Weis, K. Warkhaftig and U. Faif. 1974. *Zemlya. Vvedenie v Obshchuyu Geologiyu* (Earth. An Introduction to General Geology). Translated from English, Vol. 1, p. 378; Vol. 2, p. 753. Moscow.
16. Shvartsbakh, M. 1950. *Klimaty Proshlogo* (Climates of the Past). Translated from German. p. 84. Moscow.
17. Yuzepchuk, S.V. 1937. Novye vidy roda *Solanum* iz gruppy *Tuberarium* Dun. (New species of genus *Solanum* from the group *Tuberarium* Dun.). Izv. *An SSSR, Ser. Biol.*, No. 2, pp. 295–331.
18. Albert, F. 1900. Las dunas de Chile. Imp. Cervantes Alvaro Montapas Santiago, p. 107.
19. Anonymous. 1976. Shellfish-eating Preceramic Indians in Coastal Brazil. *Science*, No. 4225, January 30, pp. 353–360.
20. Anonymous. 1977. The cold seas of 18000 B.P. *Science News*, November 26, p. 358.
21. Atlas de la Republica de Chile. 1966. Instituto geographico militar. Santiago, p. 79.
22. Axelrod, D.I. and P.H. Raven. 1950. Evolutionary biogeography, viewed from tectonic plate. Problems and directions toward their solution. Oxford University, N.I. pp. 218–236.
23. Berger, W.H., R.F. Johnson and J.S. Killinglay. 1977. Global Freshening of Upper Ocean Layer During Deglaciation. *Naturwissenschaften*, Vol. 12, pp. 634–635.
24. Brücher, Heinz. 1966. Beitrag zur Kenntnis chilenischer Wildkartoffel. *Arten. Ber. Deutsch. Bot. Gesel.*, No. 79, Vol. 5, pp. 219–232.
25. Brücher, H. 1966. Wildkartoffeln in Afrika? *Zeitschr. für Pfl. Züchtung.* No. 56, Vol. 2, pp. 147–163.
26. Canas Pinochet Alexandro. 1901. La Papa. Investigaciones su origen, sus cultivos i las enfermedades i pestes que la atacan en Chile. *Actes Scientifique du Chile*, Vol. XI, pp. 159–197.
27. Core, L. Earl. 1950. Plant Taxonomy. U.S.A., Prentice Hall.
28. Correll, Donovan Stewart. 1962. The Potato and Its Wild Relatives. Renner, Texas, pp. 18–21.
29. Darwin, Charles. 1960. Extracts from letters addressed to Professor Henslow by Ch. Darwin, E.S.Q., read at a meeting of the Cambridge Philosophical Society, November 16, 1835. Privately printed, Cambridge, p. 31.

30. Darwin, Charles and the voyage of the Beagle. 1945. Unpublished letters and notebooks (letters No. 21, 25 to Catherine). Ed. with introduction by Nora Barlow. London, Pilot Press. pp. 100, 110, 114.
31. Fairbridge, R.W. 1977. Naturwiss. Rundschau. Vol. 8, p. 304.
32. Febrés, Andrés. 1846. Calepino de la lengua Chilena. Madrid.
33. Frakes, Lawrence and E.M. Kemp. 1972. Influence of continental positions on Early Tertiary climates. *Nature,* No. 5376. November 10.
34. Fresier, M. 1716. Ingenieur ordinaire du Roy. Relation du voyage de la mer sud aux cotes du chilly et Perou, fait pendant les année 1712, 1713, 1714, Paris.
35. Gay, Claudio. 1831. Sobre la verdadera Patria de la Papa o Patata. *El Araucano,* No. 41, June 25.
36. Gay, Claudio. 1849. Historia fisica y polytica de Chile. Botanica, Santiago, Vol. 5, pp. 74–77.
37. Gell, P., J.G. Hawkes and S.T.C. Wright. 1960. The application of immunological methods to the species within the genus *Solanum* sect. *Tuberarium. Proc. R. Soc.*, 13, 151, pp. 364–383.
38. Harschberger, John W. 1911. A.B. B.S. Ph.D. Phytogeographic Survey of North America. Die Vegetation der Erde. Leipzig.
39. Hawkes, J.G. and I. Kartoffel. 1958. Taxonomy, cytology and crossability. Handbuch Pflanzen Züchtung. Vol. 3, pp. 1–43.
40. Hawkes, J.G. 1963. A revision of the tuberbearing Solanums (second edition). Scottish Plant Breeding Station. Pentlandfield Roslin, Midlothian, p. 106.
41. Hawkes, J.G. Taxonomy, cytology and crossability (of the potato). Handbuch der Pflanzen Züchtung. Vol. 3, No. 2, pp. 1–195.
42. Hawkes, J.G. 1972. Evolution of the cultivated potato *S. tuberosum* L. *Symp. Biol. Hung.*, p. 12, pp. 183–188.
43. Herold, F. 1967. Virus investigations of potatoes from South Africa. *Am. Pot. J.*, Vol. 44, No. 1, pp. 22–23.
44. Holmes, A. 1959. A revised geological time scale. Trans. Edinburgh. *Geolog. Soc.*, 17, pp. 183–216.
45. Köppen. 1931. Grundriss der Klimakunde. Zweite verbesserte Auflage der Klimate der Erde. Berlin.
46. Lenz, Rodolfo. Dr. 1910. Diccionario Etimologico de las vose Chilenas derivadas de lenguas indijenas americans por el Dr. Rodolfo Lenz. Santiago de Chile, Palabra "Cari", p. 180.
47. Maldonado Roberto, C. 1897. Estudios geograficos e Hidrograficos sobre Chiloe. Santiago, pp. 334–339.
48. Mercer, J.H. 1976. Quickness of climate reverses. *Science News,* November 27, p. 346.
49. Molina, Juan Ignacio. 1782. Saggio sulla storia naturale del Chili. Bologna, p. 131.

50. Molina, I. 1786. Versuch einer Naturgeschichte von Chile. Leipzig, pp. 108–109.
51. Montaldo, Alvaroy Carmen Sanz. 1962. Las especies de papas silvestres y cultivadas de Chile. *Agricultura Tecnica*, Santiago, Chile, December. pp. 66–152.
52. Parkin, D.W. 1976. Solar constant during glaciation. *Nature*, March 6, pp. 28–31.
53. Pickering, Charles, M.D. 1871. Chronological History of Plants. Boston, pp. 660–662.
54. Raven, P.H. and D.I. Axelrod. 1974. Angiosperm biogeography and past continental movements. *Ann Missouri. Bot. Gard*, Vol. 61, pp. 539–673.
55. Raven, Peter H. and Daniel I. Axelrod. 1974. History of the Flora and Fauna of Latin America. pp. 420–429.
56. Reiche, Karl. 1907. Grundzüge der Pflanzenvorbreitung in Chile. Leipzig, Engelman. pp. 94, 284.
57. Reiche, C. 1910. Flora de Chile. Santiago, Vol. 5, pp. 354–355.
58. Fitzroy, Robert. 1839. Narrative of the surveying voyage of His Majesty's ships *Adventure* and *Beagle* between the years 1826–1836, description of their examination of the southern shores of South America, and the *Beagle's* circumnavigation of the globe. London, Vol. 1–3; Vol. 1, pp. XVII, 278, 287, 332; Vol. 2, pp. 363, 367, 374, 375, 379, 384, 387.
59. Roze, Ernest. 1898. Histoire de la Pomme de Terre. Paris, Rotschild.
60. Ruiz, H. and J. Pavon. 1794. Florae peruvianae and chilensis Prodromus. Madrid.
61. Simmonds, N.W. 1964. Factors in the evolution of the *Tuberarium* group. *J. Linn. Soc.*, No. 376, pp. 43–56.
62. Singer, Siegfried Fred. 1975. Edit. The Changing Global Environment. Boston, U.S.A.
63. Steward, Julian H. and Louis C. Faron. 1959. Native Peoples of South America. New York; Toronto; London.
64. Sykin [Zykin], A.G. 1971. Zur Frage der Abstammung und der wild wachsenden vorfahren chilenischer Kulturkartoffeln. *Z. Pflanzzücht.*, 65, pp. 1–14.
65. Valentine, D.H. (ed.). 1972. Taxonomy, Phytogeography and Evolution. L.N.Y., p. 287.
66. Van der Plank, J.E. 1946. Origin of the first European potatoes and their reaction to the length of day. *Nature*, Vol. 157, pp. 503–505.
67. Weberbauer, A. 1911. Die Pflanzenwelt der peruanischen Anden. Leipzig, pp. 301–306.
68. Week, L.G. 1947. Paleogeography of South America. Assoc. of Petroleum Geologists, Vol. 30, No. 7, pp. 1194–1241.
69. Weinmann, H. 1967. Basuto Potatoes "Lesotho". *J. Basuto Sci. Assoc. Masery.*

70. Whitehouse, David and Ruth, Thames and Hudson. Archeological Atlas of the World. London, 1975.
71. Zeven, A.C. and P.M. Zhukowskij. 1975. Dictionary of cultivated plants and their centers of diversity (excluding ornamental, forest trees and lower plants). Wageningen.

TRUDY PO PRIKLADNOI BOTANIKE, GENETIKE I SELEKTSII

1978 Vol. 62, No. 1

Native Varieties of *Solanum chilotanum* Hawk.

L.I. Kostina
Candidate of Agric. Sciences

The collection of native Chilean varieties of the potato was studied in 1967–1977. Some 360 specimens from the VIR collection were grouped into 50 varietal types based on the characters of tubers, sprouts, leaves and flowers. The collection is represented by many specimens: conv. montticum—53, conv. coraila—22, f. paucijugum—21, var. pehuenchense—14 and others. Single specimens are also found in the collection: var. elegans, var. infectum, var. pachacona, var. pichuna, f. acuminatum and f. camota.

An analysis of a number of Chilean forms, carried out by studying the progeny resulting from self-pollination, contributed to the establishment of the origin of some of these.

Tetraploid species of the potato *Solanum chilotanum* Hawk. are represented by a large number of botanic varieties. These forms and varieties can be distinguished from each other by a number of morphological and physiological characters. This species evidently emerged as a result of the participation of *S. molinae*, *S. leptostigma*, *S. zykinii* and their hybrids. Probably, other Chilean species of the potato were also involved in the origin of *S. chilotanum* but these have not yet been discovered or are extinct.

In Chile native varieties of the potato are disappearing fast, hence they deserve careful study. In order to solve the problems relating to the origin of the European varietal assortment of this crop, and to establish links between the native and improved European varieties, it is essential to have a preliminary systematization of the entire diversity of the Chilean potato. A study of the local varietal names is of great importance in solving the problems of evolution and systematics of the cultivated forms.

In the present work we present the result of a study made during 1967–1977 of the collection of native Chilean varieties.[1] This collection comprises specimens collected by three Soviet expeditions headed by: S.V. Yuzepchuk in 1926, P.M. Zhukovskii in 1958 and A.G. Zykin in 1967, as well as specimens supplied from England in 1964 (Commonwealth Potato Collection). Material collected by these expeditions represents a unique group, sharply different from the hilly Andean potatoes.

The endemicity of the old Chilean varieties is confirmed by the presence of undesirable characters: irregular shape of the tubers, very dark, in fact almost black, color of the tubers and sometimes of their pith, and long stolons. The diversity of the existing Chilean varieties, which are quite unique and sharply different from the European improved varieties, indicates that *Solanum chilotanum* is no doubt a native species of the potato [2, 6, 9, 15].

An analysis of the descriptions of old improved varieties shows that some Chilean local varieties were brought directly under cultivation in some countries of Europe and the USSR. Through a comparison of the descriptions of old French varieties given by Diehl in 1938 [17], with the Chilean forms present in the collection of the VIR (All-Union Institute of Plant Industry), it has been established that variety Saucisse is identical with conv. *coraila*, and Vitelotte Noire with var. *infectum*. Violette Dauvergne is also close to the old Chilean potato variety Pilinicum, which was cultivated in the '20s in Moscow province [1] and was identical to f. *pillicume*. Apart from this, improved varieties like Zolotype glazki, Pichblow and Liebscher [2] are very close to var. *chilotanum* variety Franceza negra is identical with var. *yuthense* [11] and variety Negra is close to var. *infectum*. A local variety from Czechoslovakia, Sinyukha, and a local variety of the USSR, Rozovyi from Milet, are close to *S. chilotanum*.

In the collections of S.V. Yuzepchuk, P.M. Zhukovskii and A.G. Zykin, specimens have been isolated which are identical to European varieties such as Early Rose, Magnum bonum, Blaue Riesen, Up-to-date and Wohltmann [11]. At present better Phytophthora-resistant varieties from Holland and Germany have been introduced into Chile. In 1969 Contreras [16] studied 264 specimens from Chile at the experimental station in Karilanka. Out of these 54 were found to be man-made improved varieties, of which the following had been identified: Ackersegen, Aquila, Benedetta (Benedicta), Cunca (local

[1]Material studied and systematized in this work is not included in the volume: "*Kartofel*" of the "*Kul'turnaya Flora SSSR*" (1971).

Chilean variety), Eva, Gineke, Irene, Maritta, Pimpernell and Sedafen. From material collected in Chile, A. Montaldo identified the following varieties: Green Mountain, Katahdin, Majestic, Royal Kidney and Pimpernell.

A few incidental local Chilean varieties, which formed the material for the varietal assortment of this crop in the USA and which were also used to develop many European varieties, are being used in potato breeding even today. The latter ones inherited their general plant form and valuable qualities (yield potential, early maturity, resistance to some viruses and many others). An early maturing variety, Early Rose, is widely used in breeding programs and was developed by Brezi in the USA in 1861 through selection of plants obtained from self-pollination of Garnet Chile. Even at the beginning of the '20s, 217 varieties in the USA [5] and more than 30 varieties in the USSR had been developed on the base of Early Rose.

Native varieties of the potato in Chile possess a number of desirable traits: good shape and large size of tubers, small eyes, yield potential, high protein and starch contents, high table qualities, resistance to mechanical damage, short vegetative period and neutral reaction to length of day, thus providing an opportunity to obtain a satisfactory yield under northern conditions [2, 4, 6, 10]. The collection of native forms is of special interest for breeding because of their resistance to viruses. A form resistant to X-virus (var. *villaroella*) has been identified among them and with its involvement the first varieties resistant to this virus, viz. Tawa and Saco, have been developed.

In respect to the boldness of the tubers many forms (conv. *coraila*, conv. *montticum*, f. *camota*, f. *paucijugum*, var. *pachacona* and others) are not inferior to the improved varieties. The shape of the tubers is quite varied, but an irregular shape with deep eyes is often characteristic of them. Var. *pachacona*, var. *pehuenchense* and var. *pichuna* have long horn-like tubers, deepened traces of stolons and, in the majority of cases strong anthocyanin coloration of the tubers and even of their pith. For the native forms under study it was characteristic to have good foliage and strong anthocyanin coloration of the stem and big leaves with large, wide lobes. The terminal lobe was generally ovate and bigger than the lateral ones. Dominance of sterility was characteristic for the majority of the forms, but some of them (conv. *multibaccatum*, var. *pichuna*, var. *pehuenchense* and others) develop berries.

In potato breeding the importance of the collection of native varieties increases considerably when work is undertaken to improve the health of this collection by the method of meristem culture.

SYSTEMATICS OF SOLANUM CHILOTANUM

Clones of forms of the Chilean potato are combined in groups with a complex of morphologically identical characters. The first such system was developed by S.M. Bukasov in 1933 [2]. Its main characters were: thickness of the style, its shape, width (thickness) of filaments, their position in relation to the axis of the flower and others. The entire diversity of potatoes from Chile was subdivided into five groups and two orders (*S. leptostigma* and *S. molinae*) which include 41 botanic varieties and forms [11]. The system developed by V.S. Lekhnovich includes groups of varieties and variety-forms. He identified 13 more or less distinct groups which included 59 forms and botanic varieties [11].

We systematized the collection of Chilean native varieties (370 specimens) into 51 varietal types. Initially the collection was grouped according to characters of the tuber and aerial shoots, but later on also on the basis of the characters of bush, stem, leaf and flower. The description of the plants was compiled according to the method developed in the division of tuberous plants at the VIR (3, 7–9, 12, 13). The affinity of some of the identified varietal types to corresponding forms was determined as per V.S. Lekhnovich [11]. The frequency of occurrence of specimens in distinguishable varietal types is as below:

Varietal type	*Number of specimens*	*Varietal type*	*Number of specimens*
Pachacona	5	Castro	3
Michune	14	(red-purple)	
Pichuna	8	Carribana	1
Infectum	2	Garnet Chile	7
Huicana	1	Negra	8
(blue-purple)		Chapiquina	33
Huicana	2	Frutilla	1
(red-purple)		Chamizuda	21
Frutilla	1	Chaitenera	7
Dorma	1	Clavela	17
Pechuenche	1	Coraila	22
Cedila	5	Mantequilla	15
Castro	3	Guacha	3
(blue-purple)		La Tayo	5
Bolera	4	Villjaroella	3
Multibaccatum	4	Pillan	1
Pigmentatum	2	Volcan	2
Camota	6	Quila	1
Mago	2	Magelanes	15
Franceza negra	1	Huinco	2
Gentiles	2	Pichana	4
Cielo	9	Palmeta	12
Morada chacao	1	Amarilla redonda	1
Azul	17	Roja redonda	4

Natalina	7	Afrechuda	1
Molejona	3	F. acuminatum	1
Morada ojuda	3	Var. elegans	1
Early Rose	3		

As a result of the systematization of specimens in the collection, identical specimens have been identified which makes it much easier to maintain the collection and to evaluate its biochemical properties and resistance to diseases and pests.

As a result of the grouping of the collection and the analysis of the thus identified varietal types, we could single out 45 varietal types (forms) requiring development of resistance to pests, because they were represented in the collection by solitary specimens and were highly infested with viruses.

The following were represented by a large number of specimens in the collection: conv. *montticum*—43 specimens, conv. *coraila*—25, f. *paucijugum*—22, var. *pehuenchense*—14. Those represented by only one specimen were: var. *elegans*, var. *infectum*, var. *pachacona*, var. *pichuna*, f. *acuminatum*, f. *camota*, f. *ovatum*, f. *mago*.

Conv. *montticum*, conv. *coraila* and f. *paucijugum* are less infested with viruses and therefore, a large number of their specimens survived in the collection. Conv. *coraila* was earlier the most widespread group of varieties in Chile, but it is now disappearing. One specimen of this form has been isolated from the VIR collection and has been released for cultivation in the Samarkand and Tashkent provinces under the name Abidov 2. *F. pichuna* is quite unique: it has horn-like, bent, dark-purple (almost black) tubers. *F. pehuenchense* with red, horn-like, bent tubers forms berries. A large number of specimens of f. *elongatum* are present in the collection; it flowers profusely and forms berries. *F. infectum* has oval, dark-blue-purple (almost black) tubers; the pith of the tubers is often colored with anthocyanin; the color of the corolla is white, with a blue-purple star. *F. huinco* has stone-pine shaped yellow tubers, with a light pink color at the stolon traces. The bush is tall, and vigorous. The pedicels thick and very long. It must be mentioned that in the field progeny of plants closer to this form, we did not notice anything indicating an isolated origin of this species from a species which has not yet been discovered or has become extinct [11].

An analysis of the progeny of Chilean forms from self-pollination gave us an opportunity to establish the origin of some of them.

The progeny of f. *pehuenchense* (Chalupa, k-6153; Chacho, k-7581) were comparatively homozygous regarding the morphological characters of the bush, leaf, flower and tubers. All plants (Chalupa, k-6153) had red horn-like tubers.

The progeny of conv. *multibaccatum* were quite heterogenous; there were plants closer to the original form as well as to f. *elongatum*. This confirms the opinion of S.M. Bukasov [2] that f. *elongatum* is a hybrid of v. *multibaccatum* and f. *ovatum*.

In the progeny of f. *pichuna* (k-7592, 7595, 7596) the majority of plants were closer to the initial form in regard to the morphological characters of the bush, leaf, flower and tuber with dark-violet (almost black) long, horn-like tubers. Tubers have been isolated closer to f. *pehuenchense* in respect to morphological characters; they had red horn-like tubers (k-7592, sapling 8; k-7592, sapling 10 and others), as well as some saplings closer to f. *infectum* (k-7592, sapling 6) and f. *elongati-petiolatum* (k-7592, sapling 17). The data so obtained indicate the possible origin of the above-mentioned forms from f. *pichuna* as a result of natural self-sowing or resowing by the local population.

A. Montaldo, in a letter to S.M. Bukasov in 1964, informed him that the local Chilean population in Chiloe and Languohue and German settlers sowed seeds from self-pollination which carried selection in them; they also selected plants from natural self-sown plants and maintained them subsequently. Variety Pillan is a clonal selection of potato which was grown in Puerto-Oktai. The variety Cunca, with a yellow pith and red tubers, represents plants from self-pollination of the old variety President Kruger.

Varietal type Pachacona (var. *pachacona* Lechn.). Type Yuz. 8977, horn-shaped. Tubers horn-shaped, yellow. Eyes small, pale-reddish purple. Brows long, not always distinct. Pith bright-yellow (Fig. 1, b). Aerial shoots red-purple. Base of the shoots conical, long; internode long, narrow. Apex small, slightly open. Scaly leaflet lanceolate. Pubescence dense, of average length, projecting.

The bush is of average height. Stem colored with dots of anthocyanin. Wings of average width, undulated. Leaf slightly dissected, with lobes apart from each other. Terminal lobe ovate with cordate base. Lateral lobes small, narrow-oval, with highly acute apex. Lobe stalks short.

Flowering poor as the buds often fall off. Pedicels thin, round, slightly colored with dots of anthocyanin. Articulation (joint) in the middle of pedicel, colored with anthocyanin. Pedicels thickened towards calyx and cannot be distinguished from it. Pedicels and calyx often pubescent with short pressed hairs. Acute terminal parts of calyx long, lanceolate. Corolla small, bright red-purple, with white acumens, average or highly dissected. Antheral column slightly conical, almost cylindrical. Anthers yellow. Filaments long. Style long, significantly exceeding the length of antheral column (up to 4 mm). Stigma bi- or trifid. Berries form very rarely.

The following are related to this varietal type: Yuz. 8914, horn-shaped; Yuz. 8977, horn-shaped; k-3404 a; Colique (k-6000); Dorma (k-6107).

Varietal type Michune (var. *pehuenchense* Lechn.). Type Michuna rosa (k-7551). Tubers red, horn-shaped. Eyes small. Pith cream colored, often with anthocyanin venation (see Fig. 1a). Aerial shoots dark-red-purple. Base of the shoots conical, apex closed, neck long, thin. Scaly leaflet lanceolate.

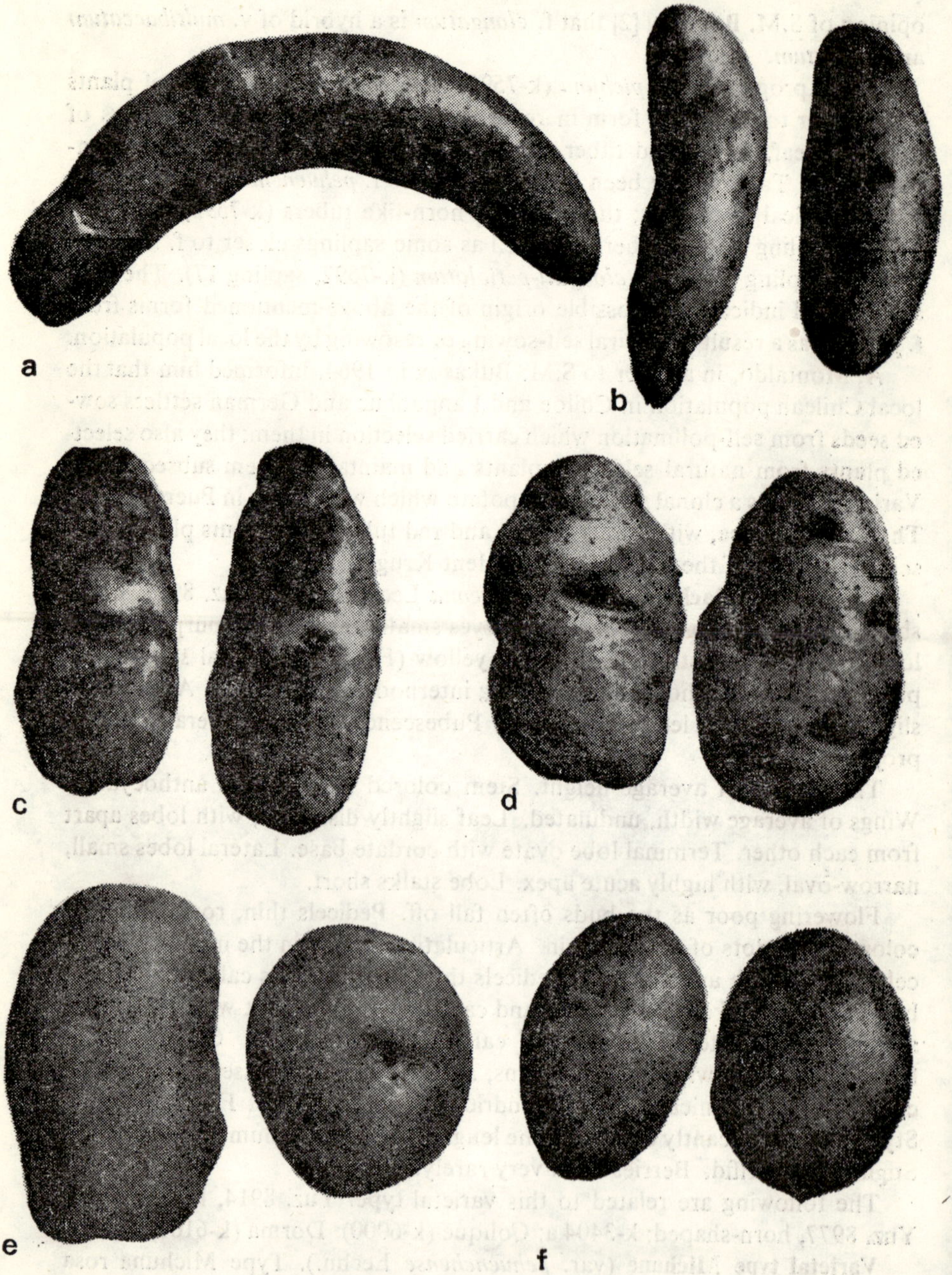

Fig. 1. Tubers of *Solanum chilotanum* Hawk.:
a—k-6008 Correto; b—Yuz. 8977 horn-shaped; c—PV-62 Franceza negra;
d—k-6108 Dorma; e—Yuz. 1980 var. villaroella; f—k-6039 Corahila pintada.

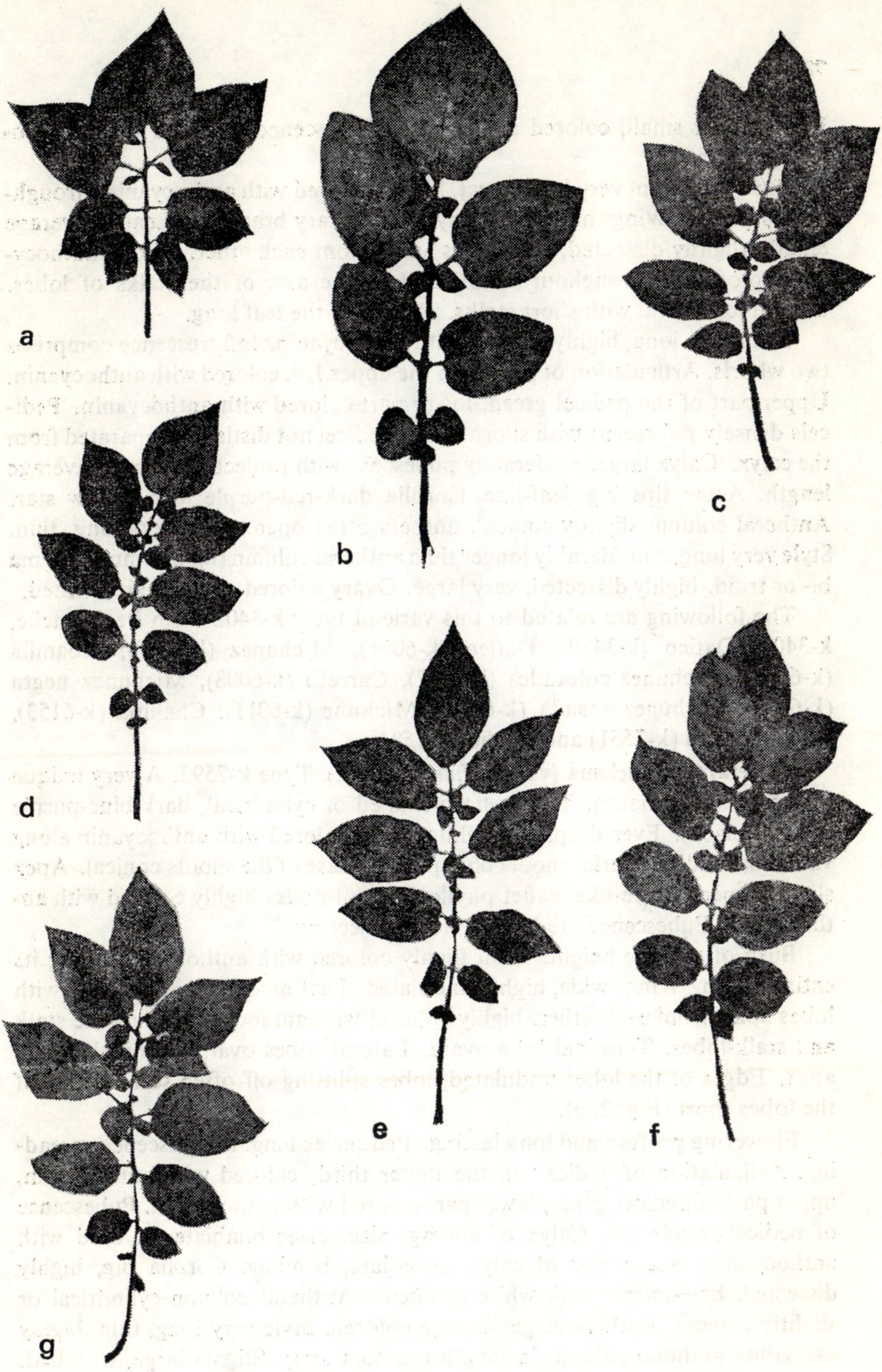

Fig. 2. *Solanum chilotanum* Hawk.:
a—k-6080 Huicana; b—k-6108 Dorma; c—k-7518 Azul de guelez; d—k-3424a Clavela 1; e—k-7546 Chamizuda; f—f. *infectum*; g—k-7595 var. *pichuna*.

Roots nodes small, colored at the base. Pubescence rare due to short projecting hairs.

Bush tall. Stem very long, erect, highly colored with anthocyanin throughout its length. Wings narrow. Many long axillary branches. Leaf of average length, slightly dissected, with lobes apart from each other. Strong anthocyanin coloration throughout petiole and in the axis of the stalks of lobes. Lateral lobes oval with short stalks. Petiole of the leaf long.

Peduncles long, highly colored with anthocyanin. Inflorescence comprises two whorls. Articulation of pedicel in the upper 1/3, colored with anthocyanin. Upper part of the pedicel green, lower part colored with anthocyanin. Pedicels densely pubescent with short hairs. Pedicel not distinctly separated from the calyx. Calyx large, moderately pubescent with projecting hairs of average length. Acute tips big, leaf-like. Corolla dark-red-purple with yellow star. Antheral column slightly conical, anthers often open. Filaments long, thin. Style very long, considerably longer than antheral column (up to 6 mm). Stigma bi- or trifid, highly dissected, very large. Ovary colored. Berries are formed.

The following are related to this varietal type: k-3402 from Pechuenche, k-3404, Datico (k-3439), Datico (k-6001), Michunez (k-6005), Jabanilla (k-6006), Michunez colorades (k-6007), Carreto (k-6008), Michunez negra (k-6009), Michunez rosada (k-6010), Michune (k-6011), Chalupa (k-6153), Michuna rosa (k-7551) and Cacho (k-7581).

Varietal type Pichuna (var. *pichuna* Lechn.). Type k-7592. A very unique group of late varieties. Tubers horn-shaped or cylindrical, dark blue-purple (almost black). Eyes deep. Pith white, often colored with anthocyanin along vascular bundles. Aerial shoots blue-purple. Base of the shoots conical. Apex slightly open. Scale-like leaflet phylloid. Root nodes highly colored with anthocyanin. Pubescence moderate, short projecting.

Bush of average height. Stem highly colored with anthocyanin along its entire length. Wings wide, highly undulated. Leaf moderately dissected, with lobes apart from each other, highly colored with anthocyanin along the stalk and stalk-lobes. Terminal lobe ovate. Lateral lobes oval with highly acute apex. Edges of the lobes undulated; lobes splitting off often seen. Stalks of the lobes short (Fig. 2, g).

Flowering profuse and long lasting. Peduncles long. Inflorescence spreading. Articulation of pedicels in the upper third, colored with anthocyanin, upper part of pedicel green, lower part colored with anthocyanin. Pubescence of pedicels moderate. Calyx of average size, often bilabiate, colored with anthocyanin. Acute tips of calyx lanceolate, bending. Corolla big, highly dissected, blue-purple with white acumens. Antheral column cylindrical or slightly conical. Anthers large, orange colored. Style very long, thin, highly exceeding antheral column in length (up to 4 mm). Stigma large, tri-lobed. Ovary colored. Berries are formed.

The following are related to this varietal type: Yuz. 1996 Pichuna, k-3407,

Majon de gato (k-3451), Michuna Azul (k-7545), k-7590, k-7592, k-7595 and k-7596.

Varietal type Infectum (var. *infectum* Buk. et Lechn.). Type PV-79 var. *infectum*. Tubers dark blue-purple (almost black), oval, slightly wrinkled. Eyes of average depth, small. Brows short, not always distinct. Stolon scar moderately depressed. Pith blue-purple with white venation. Aerial shoots dark blue-purple. Base of the shoots conical. Apex open. Scaly leaflet phylloid. Pubescence at the base absent.

Bush of average height. Stem straight, later on leaning, very prominently colored with anthocyanin. Wings of average width, slightly undulated. Axillary shoots present.

Leaf average or large, attached to stem at an acute angle. Dissection poor or average. Leaf lobes touching each other. Anthocyanin strong along petiole and particularly in the axis of the stalk. Terminal lobe bigger than lateral ones, ovate, with cordate base. Lateral lobes oval, with acute apex. Stalks of the lobes short, leaf petiole long. Wing-like structures present on the leaf stalks, decurrent along leaf stalk at the terminal lobe (Fig. 2, f).

Peduncle long, round, slightly colored with anthocyanin. Articulation of pedicel in the upper half, upper part of pedicel green, lower part colored with anthocyanin. Pedicels moderately pubescent with short pressed hairs. Articulation of pedicels colored. Upper part of pedicel thickened and not distinctly separated from the calyx. Calyx small, green, moderately or rarely pubescent. Acute tips of calyx small, lanceolate, bent. Corolla white with blue-purple star, moderately dissected (Fig. 3, d). Antheral column slightly conical. Anthers orange with blue-purple tinge in the inner side. Filaments long, yellow. Style long, thin, sometimes slightly bent, exceeds antheral column in length by up to 4 mm. Stigma capitate. Ovary colored. Berries are formed.

Pv 79 and v-170 var. *infectum* are related to this varietal type.

Varietal type Huicana (blue-purple). Morphologically close to f. *elongatipetiolatum* Lechn. Type Huicana (k-6080). The name Huicana, when translated, means not flowering. Tubers round, dark blue-purple (almost black). Eyes of average depth; brows long, distinct. Stolon scar depressed. Pith pale yellow with colored vascular ring. Aerial shoots dark blue-purple. Base oval. Apex small, open. Scaly leaflet phylloid. Root nodes average, dark with some tinge at the base. Pubescence with dense, short, projecting hairs.

Bush short-statured. Stem short, leaning, thin, highly colored with anthocyanin throughout the entire length. Wings narrow, slightly undulated. Many long axillary branches. Leaf short. Dissection poor (3 K_1l_1), but with basiscopic lobelets. Very strong anthocyanin along leaf petiole, stalk and veins of the lobes. Terminal lobe large, ovate, with cordate base. Lateral lobes average, oval touching each other. Stalks of the lobes short, leaf petiole long. Lobelets very small, despite very poor dissection of the leaf-acroscopic (Fig. 2, a). Does not flower.

Fig. 3. Corolla of *Solanum chilotanum* Hawk.:
a—Yuz. 1980 var. *villaroella*; b—k-7549 Azul; c—k-7578 Muezca; d—sapling f. *infectum*; e—k-6005 Michunez; f—k-3400 Papa de los gentiles; g—k-6100 varietal type Chapiquina.

This varietal type is represented by only Huicana (k-6080).

Varietal type Huicana (red-purple). Closer to conv. *Pyriconicum* f. *montanara* Buk. et Lechn. Type Huicana (k-6045). Tubers red (beneath the skin), wrinkled, napiform. Eyes very deep, intensively colored. Stolon scar highly depressed. Pith dark yellow.

Bush of average height. Stem moderately colored with anthocyanin. Stem nodes well manifested. Wings of average width. Leaf of average size, narrow, highly (or moderately) dissected. Terminal lobe oval with cordate base. Lateral lobes oval with very undulating margins. Apex of the lobes very acute. Anthocyanin coloration poor along leaf petioles and in the axis of the petioles. Petioles short (or medium in length). Leaf stalk of average length. Lobes are similar to those of the main lobes, with undulating margins.

Peduncle thin, highly flattened, almost without anthocyanin. Inflorescence compact. Pedicels green. Articulation of pedicels in the upper one third, colored with anthocyanin. Upper part of pedicel thick, not distinctly separated from the calyx. Pedicels pubescent with dense oblique hairs. Calyx green, densely pubescent with short oblique hairs. Acute tips of calyx lanceolate or phylloid. Corolla small, pale red-purple. Acumens small. Antheral column regular, slightly conical. Anthers small, bright orange colored. Style long, considerably exceeding antheral column (up to 5 mm). Stigma large, bifid, with deep sinus. Ovary colored. Berries not formed. Huicana (k-6045), Reina redonda (k-6063) are related to it.

Varietal type Frutilla. Type Frutilla (k-7579). Tubers variegated, red-purple, with white eyes. Eyes deep. Stolon scar highly depressed. Pith yellow.

Bush tall. Stem erect, highly colored with anthocyanin. Wings narrow, straight. Many axillary branches. Leaf highly dissected, moderately colored with anthocyanin along petiole and lobe stalks. Terminal lobe oval, with cordate base. Lateral lobes oval with sharp venation and highly acute apex. Lobe margins undulated. Stalks of the lobes long. Leaf petiole of average length.

Flowering poor. Peduncle long, thin, slightly colored with anthocyanin. Articulation of pedicels in the upper half, not colored. Calyx small, slightly colored with anthocyanin. Acute tips of the calyx lanceolate. Corolla pale red-purple (marble color). Antheral column conical. Anthers orange colored. Style very long, considerably exceeding the antheral column (up to 4 mm). Stigma bifid. Ovary colored. Frutilla (k-7579) is related to this varietal type.

Varietal type Dorma. Type Dorma (k-6108). This is a unique varietal type. Tubers round, wrinkled, bicolored: red-purple, while the upper tuber and eyes are dark blue-purple. Eyes deep. Stolon scar highly depressed, pith white (Fig. 1, d). Aerial shoots blue-purple. Base of the shoot conical. Apex small, closed. Scaly leaflet subulate. Root nodes small. Pubescent with dense, short, projecting hairs.

Bush of average height. Stem erect, colored with anthocyanin spots throughout its length. Wings wide, undulated. Many long axillary branches. Leaf average, slightly dissected, with leaflets touching each other. Anthocyanin strong along petiole and stalks of the lobes. Terminal lobe large, ovate, with cordate base. Lateral lobes large, oval. Stalks of the lobes short, leaf petiole long. Lobelets small (Fig. 2, b). Does not flower.

Dorma (k-6108) is related to this varietal type.

Varietal type Pehuenche. Type Pehuenche (k-6004). Tubers dark blue-purple (beneath the skin), round-oval or oval. Eyes of average depth or shallow; brows long, well developed. Pith bright-yellow. Aerial shoots blue-purple, develop very slowly. Base of the shoots spherical. Apex open. Root nodes average. Pubescence moderate, with short hairs.

Bush of average height. Stem leaning, slightly colored with anthocyanin. Leaf of average length, slightly dissected, with lobes touching each other. Heavy anthocyanin coloration at the axis of leaf petioles. Terminal lobe ovate, with cordate base. Lateral lobes wide-oval. Stalks of the lobes short. Leaf petiole long. Leaflets small, their shape different from that of the lateral lobes.

Flowering poor. Buds often fall off. Pedicels green, pubescent with dense oblique hairs of average length. Articulation in the upper one-third, colored with anthocyanin. Calyx densely pubescent with oblique hairs of average to considerable length. Acute tips of calyx phylloid. Corolla pale blue-purple. Anthers small. Filaments long. Style long, thin. Ovary not colored.

Only Pehuenche (k-6004) is related to this varietal type.

Varietal type Cedila. Morphologically close to f. *ovatum* Buk. et Lechn. Type Blanca de Valdivia (k-6021). Tubers round or round-oval, blue-purple (in the skin and beneath the skin). Eyes very deep, colored more intensively than skin. Stolon scar highly depressed. Pith white. Aerial shoots dark blue-purple. Base of the shoots spherical. Apex open. Scaly leaflet phylloid. Pubescence moderate with short projecting hairs.

Bush tall. Stem erect, highly colored with anthocyanin, particularly in the axis of the leaf petioles. Many axillary branches. Nodes of the stem well developed. Leaf of average size, slightly dissected, with lobes apart or touching each other. Terminal lobe ovate, with cordate base. Lateral lobes oval with acute apex and basiscopic decurrent. First pair of lobes leans slightly over the terminal lobe.

Peduncle long, thin, highly colored with anthocyanin. Branches of peduncle long, come off at an acute angle. Pubescence of pedicels moderate, with short hairs. Articulation of pedicels in the upper one-third, colored with anthocyanin. Upper part of pedicel green; lower part colored with anthocyanin. Calyx small, colored with anthocyanin, with small lanceolate acute tips. Corolla dark blue-purple (marble color). Antheral column slightly conical, almost cylindrical. Anthers big, orange colored. Filaments of average length, thick, colored with anthocyanin at the base. Style long, thick, straight, highly exceeds antheral column (up to 4 mm). Stigma narrow, with deep sinus. Ovary not colored.

The following are related to this varietal type: Cedila (k-6012), Blanca de Valdivia (k-6021), k-6076 sapling 6, V-139, V-229 (pith of the tuber yellow).

Varietal type Castro (blue-purple). Type k-3407 a. Tubers blue-purple, oval. Eyes shallow, colored more intensively. Brows indistinct. Pith white. Aerial shoots dark blue-purple. Base of the shoots conical. Apex small, slightly open. Scaly leaflet lanceolate. Pubescence with dense hairs of average length.

Stem of average length, thin, highly colored with anthocyanin. Wings very narrow; nodes of the stem well developed. Axillary branches many. Leaf of average size, slightly dissected, dark green, dull, with clear venation. Anthocyanin heavy along leaf petiole, particularly in the axis of the stalks. Terminal lobe big, ovate. Lateral lobes oval with strong basiscopic decurrent. Stalks of the lobes short. Leaflets small. Wing-like structures noticed along stalk of the leaf.

Flowering scanty; buds often fall off. Peduncle short, thin, colored with anthocyanin. Articulation of pedicels in the upper half, colored with anthocyanin. Pedicels pubescent with dense short hairs. Calyx small, deep, colored with anthocyanin, densely pubescent with short hairs. Acute tips of calyx small, lanceolate or subulate. Corolla blue-purple, slightly dissected, with wide lobes. Antheral column slightly conical almost cylindrical. Anthers orange colored. Filaments long, white, but colored with anthocyanin at the base. Style thin, straight. Stigma capitulate.

Yuz. 8874 v. *quila* f. *thalassinum* k-3407a from Castro is related to this varietal type.

Varietal type Bolera. Type Morada (k-6052). Tubers oval, dark blue-purple, color more intense at the apex of the tuber and in the eyes. Eyes of average depth. Pith bright cream colored. Aerial shoots dark blue-purple. Base of the shoots conical. Apex small, semi-closed. Scaly leaflet lanceolate. Pubescence moderate.

Bush tall. Stem erect, afterwards leaning, highly colored with anthocyanin. Wings narrow, almost erect. Many long axillary branches. Leaf of average length, slightly dissected, with lobes touching each other, highly colored with anthocyanin along leaf petiole, and particularly in the axis of the stalks of the lobes. Conical lobe of average size, ovate with cordate base. Lobes oval, dull because of dense pubescence. Stalks of the lobes short. Petiole long.

Peduncle long, highly colored with anthocyanin. Upper part of pedicel green, but lower part colored with anthocyanin. Articulation of pedicel colored. Pubescence on pedicel dense. Calyx small, highly colored with anthocyanin, moderately pubescent. Acute tips of calyx small, lanceolate. Corolla of average size, blue, with wide lobes. Antheral column slightly conical, anthers small. Filaments average or long. Style of average length; stigma bifid, dark. Ovary not colored.

The following are related to this varietal type: Yuz. 8874, k-3407a from Castro, Morada (k-6052), Bolera (k-7553).

Varietal type Multibaccatum (conv. *multibaccatum* var. *bastoneza* Buk. et Lechn.). Type V-139. Tubers oval, blue-purple. Eyes shallow. Pith cream colored with anthocyanin veins. Aerial shoots dark blue-purple. Base of the shoots spherical. Apex open. Scaly leaflet phylloid or lanceolate; root nodes not colored with anthocyanin. Pubescence moderate with short and projecting hairs.

Bush of average height. In the beginning stem erect, afterwards leaning, highly colored with anthocyanin. Wings average or narrow undulated. Axillary branches present. Leaf highly dissected, with lobes touching each other or apart from each other. Intense anthocyanin coloration along leaf petiole and particularly in the axis of the stalks of the lobes. Terminal lobe ovate; lateral lobes narrow-oval, with highly acute apex. Petiole of average length. Leaflets big; their shape similar to that of lateral lobes.

Flowering profuse. Peduncle very long, colored with anthocyanin. Pedicels thick, highly flattened towards calyx. Articulation of pedicel in the upper half, colored with anthocyanin. Pedicel moderately pubescent with short pressed hairs. Calyx small, often bilabiate, colored with anthocyanin. Acute tips of calyx of average length, lanceolate. Corolla dark-blue purple (marble colored), moderately dissected. Antheral column slightly conical, regular. Anthers large, wide, orange colored. Filaments thick. Style long, erect. Stigma with deep cone. Ovary colored with anthocyanin. Flowering in two whorls. Berries are formed.

The following are related to this varietal type: V-141, V-142, V-143 and V-288.

Varietal type Pigmentatum (conv. *chilotanum* f. *americana pigmentatum* Lechn.). Type PV-147 *pigmentatum*. Tubers dark-blue (variegated), napiform, wrinkled. Eyes very deep. Brows short, not always distinct. Stolon scar highly depressed. Pith cream colored. Aerial shoots blue-purple. Base of the shoots spherical. Apex small, closed. Scaly leaflet subulate. Root nodes average, colored with tinge at the base. Base moderately pubescent with projecting hairs of average length.

Bush of average height. Stem erect, highly colored with anthocyanin along the entire length. Wings undulated. Axillary branches present. Leaf slightly dissected with lobes touching each other, colored with anthocyanin along petiole and at the axis of stalks of the lobes. Terminal lobe ovate, with cordate base. Lateral lobes oval, with basiscopic decurrent. Stalks of the lobes short. Leaflets round, their shape different from that of lateral lobes.

Flowering profuse. Peduncles long, highly colored with anthocyanin. Pedicels also colored with anthocyanin. Articulation of pedicels in the upper half, highly colored with anthocyanin. Pedicels moderately pubescent with pressed hairs. Calyx of average size, colored with anthocyanin, often bilabiate. Acute tips of calyx phylloid. Corolla blue-purple, with bright shaded spots on the acute tips, lobes white on the lower side. Antheral column cylindrical. Anthers narrow, straight. Filaments long, white, thin. Styles long, bent. Stigma capitulate. Ovary colored.

PV-147 *pigmentatum* and PV-163 are related to this type varietal type.

Varietal type Camota. Morphologically closer to conv. *chilotanum* f. *camota* Buk. et Lechn. Type Camota (k-7552). Tubers round, variegated, blue-purple, with white color near the eyes. Eyes deep. Trace of stolon highly inserted. Mesocarp creamish in color. Aerial shoots dark blue-purple. Base of the shoots spherical. Apex small, closed. Scaly leaflet subulate. Base moderately pubescent with projecting hairs of average length.

Bush of average height. Stem thin, highly colored with anthocyanin. Wings of average width, undulated. Nodes of the stem well developed. Axillary branches present. Leaf moderately dissected, with lobes touching each other, with intense anthocyanin coloration along the petiole and in the axis of the stalks of the lobes. Terminal lobe oval, with intermediate type of base. Lateral lobes narrow-oval, with highly acute apex and sharp venation. Stalks of the lobes short.

Flowering poor, buds often fall off. Peduncle short, colored with anthocyanin. Articulation of pedicels colored (in contrast with f. *camota* Lechn.). Pedicels moderately pubescent with short pressed hairs. Calyx small, deep, slightly colored with anthocyanin, moderately pubescent with short pressed hairs. Acute tips of calyx phylloid. Corolla small, pale blue-purple, highly dissected. Antheral column deformed. Anthers not big, yellow. Filaments short, colored with anthocyanin at the base. Style straight. Stigma bifid. Ovary colored.

The following are related to this varietal type: a) articulation of pedicels colored [Camota, k-7552; Camota; V-902; PV-115]; b) articulation of pedicels not colored (Yuz. 8844, Yuz. 8864, Yuz. 8920, V-86).

Varietal type Mago (f. *mago* Lechn.). Type PV-152. Tubers short, napiform, wrinkled, dark blue-purple. Eyes deep. Stolon scar highly depressed. Pith yellow. Aerial shoots dark blue-purple. Base of the shoots spherical-oval. Apex small, slightly open. Scaly leaflet lanceolate. Pubescence at the base due to rare, short, projecting hairs.

Bush of average height. Stem erect, intensely colored with anthocyanin (almost brown) throughout its length. Wings narrow, slightly undulated. Axillary branches absent.

Leaf of average length, narrow, moderately bissected, with lobes apart from each other. Anthocyanin very intense throughout the petiole, particularly at the axis of the stalks of the lobes. Terminal lobe small, ovate, with cordate base. Lateral lobes small, oval, with highly acute apex. Stalks of the lobes short, leaf petiole of average length.

Flowering poor. Buds often fall off. Articulation of pedicels colored. Upper part of pedicels green. Corolla dark blue-purple.

PV-152 and V-234 are related to this varietal type.

Varietal type Franceza negra (conv. *glaucum* var. *yutuense* Lechn.). Type PV-210 Franceza negra. Tubers with eyes having a white ring around them, yellow, with blue-purple spots (see Fig. 1, c). Eyes of average depth. Pith yellow. Aerial shoots blue-purple. Bush of average height. Stem thin, erect, highly colored with anthocyanin. Wings narrow, undulated. Axillary branches absent. Leaf small, slightly dissected, with lobes touching each other. Anthocyanin average along petiole and stalks of the lobes. Terminal lobe oval, with cordate base. Lateral lobes oval. Stalks of the lobes short. Leaf petiole long.

Flowering moderate. Articulation of pedicels in the upper one third, colored with anthocyanin. Pedicels colored with anthocyanin, pubescent with dense short hairs. Corolla blue-purple, highly dissected. Antheral tube regular. Style straight, thick. Ovary not colored with anthocyanin.

This varietal type is represented by PV-210 only.

Varietal type Gentiles. Closer to conv. *glaucum* var. *brevipilosum* Buk. et Lechn. Type Papa de los gentiles (k-3401). Tubers bright red (beneath the skin), round. Eyes deep or of average depth, colored much brighter. Brows indistinct. Pith dark yellow. Aerial shoots red-purple. Base conical. Apex open. Scaly leaflet phylloid. Root nodes small. Pubescence very dense due to short projecting hairs.

Bush of average height. Stem leaning, highly colored with anthocyanin. Wings of average width, undulated. Axillary branches few. Leaf not big, slightly dissected. Lobes small, lustrous, with acute apex and undulating margins of the lobes. Terminal lobe insignificantly bigger than lateral lobes, with cordate base. Anthocyanin on leaf petiole and stalks of the lobes. Stalks of the lobes short or of average length. Leaf petiole long. Flowering poor, buds often fall off.

Peduncle of average length, colored with anthocyanin. Pedicels pubescent with dense short pressed hairs. Articulation of pedicels in the upper half, colored with anthocyanin. Upper part of pedicel green; lower part colored with anthocyanin. Calyx small, slightly coloreh with anthocyanin, moderately (or densely) pubescent with pressed hairs. Acute tips subulate. Corolla red-purple, small (see Fig. 3, f). Antheral column small, anthers yellow. Style straight.

Papa de los gentiles (k-3401), k-3400 are related to this varietal type.

Varietal type Cielo (var. *elongatum* Lechn.). Type Mira Cielo (k-7557). Represented by numerous varieties. Tubers round, short, pale blue-purple. Eyes deep, colored more intensely than tubers. Stolon scar highly depressed. Pith white. Aerial shoots dark blue-purple. Base of the shoots spherical. Apex small, closed. Scaly leaflet subulate. Root nodes intensely colored with tinges at the base. Pubescence rare, almost absent. Stolons appear early on the shoots.

Bush of average height. Stem thick, erect, but afterwards leaning. Anthocyanin coloration throughout the entire stem and particularly at the axis of the leaf. Wings wide, undulated. Leaf long, moderately to highly dissected, with lobes touching each other, with heavy anthocyanin coloration in the axis of the stalks of the lobes. Terminal lobe oval, with cordate base. Lateral lobes oval, with undulated margins. Stalks of the leaves of average length or short. Leaf petiole short. Leaflets big.

Flowering profuse. Peduncle long, slightly colored with anthocyanin dots. Inflorescence of two to three whorls. Pedicel green, thick. Articulation colored with anthocyanin. Calyx big, not clearly separated from pedicel. Acute tips long, lanceolate. Corolla big, blue-purple, highly dissected (see Fig. 3, c). Antheral column cylindrical. Filaments long, colored with anthocyanin in the adnate part. Style long, thick. Ovary colored with anthocyanin. Berries are formed.

The following are related to this varietal type: k-5996, Cielo (k-6028), Cielo (k-6077), k-6155, Cielo (k-6169), Zapatona (k-7526), Mira Cielo (k-7557), Redonda amarilla (k-7562), Muezca (k-7578).

Varietal type Morada chacao. Type Morada chacao (k-6118). Tubers blue-purple, round. Eyes small. Pith white with anthocyanin venation. Aerial shoots blue-purple. Stem of average length, leaning, intensely colored with anthocyanin. Wings wide, undulating. Axillary branches present. Leaf of average length, slightly dissected, with lobes touching each other. Anthocyanin poor in the axis of the stalks of the lobes. Terminal lobe big, ovate, with cordate base. Lateral lobes oval with acute apex. Stalks of the lobes short. Leaf petiole long. Flowering poor. Peduncle short. Articulation of pedicels in the upper half, colored with anthocyanin. Upper part of pedicel green but lower part colored with anthocyanin. Pedicels pubescent with dense long hairs. Calyx of average size, densely pubescent with long hairs.

This varietal type is represented by only Morada chacao (k-6118).

Varietal type Azul (f. *barmacca* Buk. et Lechn.). Type Azul (k-7510). Tubers blue-purple (beneath the skin), oval or round oval. Eyes small. Pith cream colored. Aerial shoots blue-purple. Base of the shoots oval or conical. Apex slightly open. Scaly leaflet lanceolate or phylloid. Root nodes small. Pubescence dense with projecting hairs of average or more length, hairs pressed in the upper part of the base.

Bush very tall, vigorous. Stem thick, leaning, highly colored with anthocyanin along the entire length, particularly in the axis of the leaf. Many axillary branches. Leaf big, slightly dissected with imposing lobes, anthocyanin in the leaf petiole and in the axis of the stalks. Terminal lobe large, ovate. First pair of lobes lie on the terminal lobe. Lateral lobes big, oval. Strong basiscopic decurrent in the first pair of lobes. Stalks of the leaves of average length. Leaf petiole long. Leaflets big, their shape similar to that of the lateral lobes (Fig. 2, c).

Flowering moderate or poor. Peduncle of average length, colored with anthocyanin. Pedicels colored with anthocyanin and moderately pubescent with leaning hairs of average length. Articulation of pedicels in the upper half, not colored with anthocyanin. Calyx of average size, colored with anthocyanin and pubescent with pressed hairs of average length. Acute tips of corolla phylloid. Corolla big, white, with yellowish star. Acumens very short. Anthers often deformed, pale yellow. Style thick, long, curved. Stigma very large, capitulate. Ovary colored. One berry was formed in 1972.

The following are related to this varietal type: k-3406a, Azul (k-6015), Morada (k-6047), Morada lisa (k-6049), Morada (k-6050), Azules (k-6074). Rinones (k-6081), Castro (k-6113), Clavela (k-6167), Rosada (k-7508), Azul (k-7510), Azul de gueles (k-7518), Negra (k-7520), Azul (k-7550), k-7558, Cabra (k-7583), V-909 f. *barmacca*.

Varietal type Natalina. Type Natalina (k-7523). Very close to varietal type Azul in respect to morphological characters. Tubers pink, round. Eyes of average depth. Pith cream colored. Aerial shoots reddish-purple. Base conical. Apex large, open. Scaly leaflet big, phylloid. Pubescence moderate with long, projecting hairs.

Bush tall, vigorous. Stem long, erect, thick, colored with anthocyanin throughout the entire length. Wings of average width, undulated. Stem nodes well developed. Axillary branches present.

Leaf very big, moderately dissected, with slight anthocyanin coloration, leaf lobes densely pubescent. Terminal lobe bigger than lateral, with cordate base. Lateral lobes big, oval, with acroscopic uneven edges. Apex of lobes acute. Stalks of the lobes long. Leaf petiole of average length. Leaflets very big.

Flowering poor. Peduncle of average length or long, thin, densely pubescent. Inflorescence closed, branches of peduncle short, coming out at an acute angle. Pedicels colored with anthocyanin and densely pubescent with pressed hairs. Articulation of pedicels in the upper half, not colored with anthocyanin. Calyx small, colored with anthocyanin and densely pubescent. Acute tips of calyx small, lanceolate. Corolla white (Fig. 3, d). Antheral column slightly conical, often deformed. Anthers yellow. Filaments long. Style long, thick, sometimes bent. Stigma 2–3 lobed. Ovary colored. Berries not formed.

The following are related to this varietal type: k-3407 (d-1), mixture of local varieties from Castro; k-3447 from Centinella; Tercauna (k-6141); Rosada (k-7522); Natalina (k-7523); Amarilla redonda (k-7530); Azul (k-7549).

Varietal type Molejona. Type Rinona (k-7511). As regards the morphological characters of bush, leaf and flower it is close to conv. *montticum*. Tubers blue-purple (beneath the skin), oval, flat. Eyes shallow. Pith cream colored. Aerial shoots dark blue-purple. Base oval. Apex slightly open. Scaly leaflet lanceolate. Root nodes big. Pubescence dense with long, projecting hairs.

Bush of average height. Stem erect, later on leaning, highly colored with anthocyanin. Wings almost straight. Lateral branches present. Leaf of average length, slightly dissected, lobes apart from each other or touching each other. Anthocyanin along the stalk and in the axis of the stalks. Terminal lobe oval, base straight. Lateral lobes oval. Stalks short, leaf petiole of average length.

Peduncle short, densely pubescent. Buds often fall off. Articulation of pedicel in the upper 1/3, colored with anthocyanin. Calyx big, densely pubescent with long projecting hairs. Acute tips of calyx long, lanceolate. Corolla white, lobes wide. Anthers small.

Rinona (k-7511), Molejona (k-7527) and Larga (k-7529) are related to this varietal type.

Varietal type Morada ojuda. Type Morada ojuda (k-3455). Tubers dark blue-purple, big (napiform), wrinkled. Eyes deep. Brows not always distinct. Pith dark-yellow. Aerial shoots blue-purple. Base spherical. Apex small, slightly open. Scaly leaflet subulate. Pubescence very rare, almost absent.

Bush tall. Stem long, highly colored with anthocyanin throughout the entire length. Wings undulating, narrow. Leaf moderately dissected, with lobes touching each other and intense anthocyanin coloration along the leaf petiole, stalks and venation of the lobes. Terminal lobe ovate, with cordate base. Lateral lobes oval, apex of the lobes acute. Stalks of the lobes short. Leaf petiole of average length.

Flowering scanty. Inflorescence with just a few flowers. Articulation of pedicels in the upper half, colored with anthocyanin. Upper part of pedicels green, lower part colored with anthocyanin. Pedicel and calyx moderately pubescent with short hairs. Calyx colored with anthocyanin. Acute tips of calyx lanceolate. Corolla small, pale blue-purple. Antheral column conical. Anthers small, greenish-yellow. Style long, thick, sometimes bent. Stigma large, bifid. Ovary colored.

The following are related to this varietal type: Morada ojuda (k-3455), Bollo de Chancho (k-6019), Morada ojuda (k-6048).

Varietal type Early Rose. Type k-3408 from Ankud. Tubers pink, oval. Eyes shallow. Pith white with pink ring. Aerial shoots red-purple. Base of the shoots spherical-oval (or oval), densely pubescent with hairs of average length. Apex semi-open, big. Scaly leaf phylloid.

Bush short-statured, well foliated. Stem with anthocyanin pigmentation. Leaf moderately dissected, sometimes leaflets or lobelets are present. First pair of lobes leans over the terminal lobe. Lobes smooth, glabrous, with anthocyanin in the axis of stalks of the lobes.

Flowering moderate. Peduncles not long. Peduncles, pedicels and calyx with anthocyanin. Calyx densely pubescent. Acute tips of calyx long, phylloid. Corolla white, not big. Antheral column regular, anthers orange. Style straight or slightly curved at the base, exceeds the antheral column in length. Ovary colored. Berry formation observed rarely.

The following are related to this varietal type: k-3408 c from Ankud, Colorada pintada (k-6036), Rosada Caucahue (k-6092).

Varietal type Castro (red-purple). Type k-3407d from Castro. It is very close to varietal type Early Rose in respect of leaf and flower. Tubers pale pink, round. Eyes deep. Brows long, distinct. Pith bright yellow. Aerial shoots dark red-purple. Base of the shoots conical, densely pubescent with projecting hairs of average length. Apex open. Scaly leaflet phylloid.

Bush of average height. Stem moderately pubescent with anthocyanin. Axillary branches present. Leaf slightly dissected. Lobes small, with acute apex and slightly undulating margins. First pair of lobes leans over the terminal lobe. Anthocyanin poor along the petiole and in the axis of the stalks of the lobes.

Flowering moderate. Peduncle short, slightly colored with anthocyanin. Articulation of pedicels in the upper half, not colored with anthocyanin. Pubescence of pedicels dense due to short oblique hairs. Calyx moderately pubescent with short pressed hairs. Acute tips of calyx small, lanceolate. Corolla white with yellow star. Antheral tube conical, anthers yellow or orange. Filaments long. Style curved. Stigma capitulate, wide. Ovary colored. Berries formed rarely.

The following are related to this varietal type: k-3407g and k-3407d (both from Castro).

Varietal type Carribana. Type Carribana (k-6160). Tubers bright-red to pink, round, somewhat flat. Eyes of average depth. Pith white. Aerial shoots dark red-purple. Base of the shoots spherical, rarely pubescent with projecting hairs. Apex large, slightly open. Scaly leaflet phylloid. Regarding morphological characters, very close to variety Early Rose.

Flowering moderate. Peduncle thin, colored with anthocyanin. In contrast to the variety Early Rose, the upper part of pedicel green, lower part colored with anthocyanin. Pubescence of pedicels dense, hairs short, oblique. Calyx colored with anthocyanin. Corolla white with yellow star. Antheral column conical. Anthers yellow. Style long, straight, highly exceeds antheral column in length (up to 3 mm). Ovary colored. Many berries formed.

This varietal type is represented only by Carribana (k-6160).

Varietal type close to Garnet Chile. Type k-6101. Tubers pink, round-oval. Eyes of average depth, colored more intensely. Pith yellow. Aerial shoots red-purple. Base of the shoots spherical, very rarely pubescent. Apex closed, loose, not compact. Scaly leaflet subulate or lanceolate.

Bush of average height, well foliated. Stem thick, slightly colored with anthocyanin dots, with wide undulating wings. Leaf moderately dissected, with lobes touching each other. Lobes big, dull, with venation. First pair of lobes leans slightly over the terminal lobe. Stalks of the lobes short. Axis of lobe-stalks colored with anthocyanin.

Flowering poor, buds fall off. Peduncles of average length, slightly colored with anthocyanin. Inflorescence closed. Articulation of pedicels in the upper half, colored with anthocyanin. Pedicels colored with anthocyanin and densely pubescent with hairs of average length. Buds densely pubescent with long projecting hairs. Corolla pale red-purple. Anthers greenish. Style curved. Ovary not colored.

To this varietal type belong k-3488 and 3483 from Centinella, k-3510, 6044, 6101, 6105 and Frutilla (k-6079).

Varietal type Negra. Close to f. *tenuipedicellatum* Lechn. Type k-3403 from Castro.

Tubers dark blue-purple (beneath the skin), oval. Eyes deep. Brows distinct, long. Pith white or cream colored. Aerial shoots blue-purple. Base of the shoots spherical. Apex big, open. Scaly leaflet phylloid. Pubescence dense, short hairs.

Type of bush and leaf similar to that of Early Rose. Stem thin, inclined, highly colored with anthocyanin. Axillary shoots present. Wings of average size, moderately dissected with lobes touching each other. Anthocyanin coloration along the leaf petiole and particularly in the axis of the lobe-stalks. Terminal lobe ovate, with cordate base. First pair of lobes leans over the terminal lobe. Lobes oval, apex of the lobes acute lustrous, with sharp venation. Stalks of the lobes of average length. Leaf petiole long.

Flowering moderate or poor. Peduncles of average length, thin, moderately colored with dots of anthocyanin. Inflorescence compact. Pedicels slightly colored with anthocyanin and moderately pubescent with oblique hairs. Articulation in the upper 1/3, not colored with anthocyanin. Calyx of average size, colored with anthocyanin and densely pubescent with pressed hairs. Acute tips of calyx big, lanceolate or phylloid. Corolla white with yellowish star. Lobes wide, with small acumens. Antheral column slightly conical. Anthers small, orange colored. Style long, highly exceeding antheral column in length (up to 4 mm). Style large, capitulate. Ovary colored.

The following belong to this varietal type: k-3403 from the site of the Pejuenche tribe, k-3444, Negra (k-6054), Azul redonda (k-6131), Azul (k-7528), Azul (k-7532), Bastoneza (k-7580), Papa azul (k-7597).

Varietal type Chapiquina (conv. *montticum* Lechn.). Type Chapiquina (k-6041). Tubers round, bicolored, red-purple, with blue-purple or red anthocyanin at the top and eyes. Eyes of average depth, intensely colored. Stolon scar highly suppressed. Pith yellow.

Aerial shoots dark blue or red-purple. Base spherical-oval. Apex closed or slightly open. Root nodes average, not colored with anthocyanin. Scaly leaflet phylloid. Pubescence rare or moderate, with short, projecting hairs.

Bush tall. Stem long, erect or inclined, highly colored with anthocyanin. Wings wide, undulating. Nodes of the stem well developed. Axillary branches present. Leaf long, moderately or highly dissected, with lobes touching each other. Anthocyanin throughout leaf petiole and in the axis of the stalks of lobes. Terminal lobe ovate, with cordate base. Lateral lobes oval, with highly acute apex. Stalks of the lobes and leaf petiole long. Leaflets big, similar to lateral lobes. Leaves of the English ivy type are observed (Fig. 2, d).

Peduncle long, thin, colored with anthocyanin. Upper part of pedicels green, lower part colored with anthocyanin. Articulation in the upper 1/3, not colored with anthocyanin. Pubescence of pedicels moderate. Calyx of average size, colored with anthocyanin, moderately pubescent with projecting hairs. Acute tips of calyx lanceolate. Corolla white with yellow star (Fig. 3, g). Antheral column slightly conical. Anthers small, yellow. Style long, thin, highly exceeding antheral column in length (up to 3 mm). Stigma bifid. Ovary colored. In some varieties tubers red-purple; eyes, top of the tubers and shoots blue-purple: Chapiquina (k-6041), Pachani (k-6075), k-6100, Barco (k-6159), Frutilla (k-7575).

In other varieties tubers and shoots red-purple; Yuz. 8933, k-3407 (b-1, r-1, e)—mixture of local varieties from Castro, k-3408, Blanca curaco (k-3417), Blancos varies (k-3420), Clavela 1 (k-3424), Clavela 2 (k-3426), Comuy (k-3429), Frutilla (k-3442), Mundial (k-3456), k-3480, 3482, Comuy (k-6030), Frutilla (k-6069), Rosada (k-6091), k-6095, 6099, Cominante (k-6124), Valdiviana (k-6139), Clavela (k-6158), Frutilla (k-6171), k-7589, 3081, 810 and 729.

Varietal type Frutilla. Type Palmeta (k-7499). In regard to a number of morphological characters it is close to conv. *montticum* Lechn. Tubers pink, not always uniformly colored, round. Eyes of average depth, colored more intensely. Brows not distinctly expressed. Pith yellow. Aerial shoots dark red-purple. Base of the shoots oval. Apex small, closed. Scaly leaflet lanceolate. Pubescence rare due to short projecting hairs.

Bush tall. Stem erect, moderately colored with anthocyanin throughout. Wings narrow, erect. No axillary branches. Leaf small, slightly dissected, with lobes touching each other. Anthocyanin coloration poor at the axis of the lobe stalks. Terminal lobe oval with cordate base. Lobes with sharp venation, dull. Lobe stalks short.

Flowering poor; buds fall off. Inflorescence compact. Upper part of pedicel green, lower part colored with anthocyanin. Articulation of pedicels in the middle, colored with anthocyanin. Calyx large, moderately colored with anthocyanin, with lanceolate or phylloid acute tips. Corolla small, white, with spots of red-purple anthocyanin. Antheral column conical. Style straight, exceeds antheral column in length (up to 3 mm). Stigma capitulate.

This varietal type is represented by Palmeta (k-7499) only.

Varietal type Chamizuda. Close to f. *paucijugum* Lechn. Type Morada ala (k-7564).

Tubers from bright to dark red (color beneath the skin), round to oval, with somewhat constricted top. Eyes of average depth or even deeper, often colored more lightly. Tubers wrinkled. Pith bright yellow. Aerial shoots dark red-purple. Base of the shoots conical. Apex opening slightly. Scaly leaflet phylloid. Pubescence rare due to short projecting hairs. Stolons appear very early.

Bush of average height or tall. Stem erect, thin, highly colored with anthocyanin throughout its length. Wings narrow, erect. Many axillary branches. Leaf is attached to stem at an acute angle, particularly the young leaf. Leaf long, slightly or moderately dissected, with lobes quite apart from each other. Anthocyanin coloration heavy on leaf petiole, on lobe stalks and on the main veins. Terminal lobe bigger than the lateral ones, obovate, base of the lobe straight, rarely wedge-shaped. Lateral lobes oval, with highly acute apex. Stalks of the lobes short. Leaf petiole long (Fig. 2, e).

Flowering scanty. Inflorescence with a few flowers, buds often fall off. Articulation of pedicels in the upper half; upper part of pedicel green. Calyx green, with big lanceolate or phylloid acute tips. Corolla pale red-purple. Anthers small, yellow. Style often curved. Berries not formed.

The following are related to this varietal type: k-3408b, Corahila colorada (k-6032), Corahila (k-6116), Corahila mira (k-6138), Alcahuana (k-6145), Sandia (k-6150), Pichano o chamizuda (k-6168), Palmeta (k-7500), Rosada (k-7506), Corahila fina (k-7534), Rama (k-7536), Chamizuda (k-7546), Corahila rama (k-7556), Morada (k-7559), Morada alargada (k-7560), Rosada (k-7561), Roja alargada (k-7563), Morada ala (k-7564), Rosada redonda (k-7567), k-7574.

Varietal type Chaitenera. Type Rio (k-6164). Tubers round, white, with weak blue-purple anthocyanin on the top. Eyes deep, intensely colored. Pith yellow. Aerial shoots dark blue-purple. Base of the shoots cylindrical. Apex small, closed. Scaly leaflet subulate. Pubescence moderate with very short projecting hairs.

Bush of average height. Stem leaning, thin, slightly colored with anthocyanin. Wings narrow. Axillary branches many. Leaf small, slightly dissected. Lobes small, dull, dark green with bluish tinge. Terminal lobe somewhat bigger than lateral ones, with cordate base. In the first pair of lobes strong basiscopic decurrent and slight, inclination over the terminal lobe. Flat leaves often observed. Stalks of the lobes short. Corolla blue-purple, with white acute tips.

The following are related to this varietal type: Chanques (k-6123), Mijo (k-6135), Generala (k-6140), Rio (k-6144), Chaitenera (k-6146), Osarnina (k-6157), Tontas (k-6170).

Varietal type Clavela. Type k-7594. Tubers oval, variegated; red spots on bright base color which are often confined to eyes and top of the tuber (color beneath the skin). Eyes shallow, brows not always well developed. Stolon scar slightly depressed (Fig. 1, f). Pith creamish. Aerial shoots red-purple. Base of the shoots conical, very narrow. Apex small, closed. Scaly leaflet lanceolate. Root nodes small. Pubescence dense with short projecting hairs.

Bush of average height. Stem erect, later leaning, moderately colored with anthocyanin, particularly at the axis of the leaf petiole. Wings almost absent. Axillary branches numerous. Base of the axillary branches highly colored with anthocyanin. Leaf of average size, slightly dissected, with lobes touching each other. Terminal lobe big, oval, with intermediate base. Lateral lobes oval, with strong basiscopic decurrent. Lobes lustrous. Stalks of the lobes short. Leaf petiole of average length. Flat leaves often found. Does not flower.

The following are related to this varietal type: Borrega (k-3422), Clavela 3 (k-3425), Achao (k-6013), k-6019b, Clavela (k-6023), Clavelos (k-6024), Corahila pintada (k-6039), k-6085, Clavela (k-6109), Blanca (k-6132), Clavela (k-6174), Bandera (k-7502), Clavela (k-7514), Clavela (k-7519), Clavela (k-7531), Clavela alargada (k-7569) and k-7594.

Varietal type Coraila (conv. *coraila* Lechn.). Type k-6097. Tubers oval, red, pink or yellow. Eyes of average depth. Pith yellow. Aerial shoots red-purple. Base of the shoots conical, apex big, slightly opening. Scaly leaflet phylloid. Pubescence very rare, sometimes absolutely absent.

Bush of average height. Stem leaning, colored with anthocyanin, with narrow undulating wings and large number of axillary branches. Leaf small, slightly or moderately dissected, with lobes touching each other. Terminal lobe ovate, with cordate base; lateral lobes with highly acute apex. Stalks of the leaflets of average length, leaf petiole long.

Inflorescence often deformed, branches of peduncle sprout into shoots. Peduncle long. Inflorescence of 2–3 whorls, flowers few, buds often fall off. Articulation of pedicels in the upper 1/4, colored with anthocyanin. Pedicels slightly colored with anthocyanin, moderately pubescent, somewhat expanded towards the top. Calyx large, colored with anthocyanin, with big phylloid acute tips. Corolla bright red-purple, highly dissected. Antheral tube often deformed. Style thick, curved, sometimes fused together with many styles. Stigma large, bifid. Berries not formed.

The following are related to this varietal type: k-3408c from Ankud, Coraila blanca (k-3431), Coraila larga (k-3434), k-3478, k-3479, k-3495, from Centinella, Clavela (k-6022), Mantecosa (k-6027), Colorada (k-6028), Coraila de Castro (k-6031), Coraila macho (k-6035), Coraila larga (k-6038), Pichana (k-6057), Reina (k-6062), Coraila blanca (k-6067), k-6069, 6097, Coraila larga (k-6106), k-6114 (from Castro), Coraila lisa (k-6129), V-1093, Abidov 2. The characters of this varietal type are dominant in the hybrids k-3499 (Coraila × Mohawk) and k-3503 (Coraila × Mohawk).

Varietal type Mantequilla. Type Mantequilla (k-7576). A large group of varieties. Tubers round, wrinkled, yellow, with blue-purple anthocyanin at the top. Eyes deep. Stolon scar highly depressed. Pith yellow. Aerial shoots (haulms) dark blue-purple. Base of the shoots spherical, apex small, slightly open; scaly leaflet lanceolate. Pubescence moderate (or rare) with very short projecting hairs.

Stem long or of average length; in the beginning erect but afterwards leaning, thin, highly colored with anthocyanin. Nodes of the stem green, well developed. Wings narrow, undulating. Leaf of average length, slightly or moderately dissected, with lobes touching each other. Weak anthocyanin coloration on the petiole and at the axis of stalks of the lobes. Terminal lobe oval, with cordate base; lateral lobes oval. Leaflets small, their shape different from that of lateral lobes, sessile. Stalks of the lobes and petiole of average length.

Flowering profuse. Peduncle long, with 2–4 whorls, many flowers. Articulation of pedicel in the upper 1/4, not colored with anthocyanin. Corolla large, pale blue. Antheral column conical, style long. Many berries formed.

The following are related to this varietal type: Amarilla temprana No. 2 (k-3413), Amarilla temprana No. 1 (k-3414), Amarilla temprana (k-6014), Mantequilla (k-6088), Cana negra (k-6076), Natalina (k-6110), k-6103, Dos veces al ano (k-6164), Rama (k-6155), Alcachea (k-7509), Clavela (k-7515), Anibal (k-7539), Mantequilla (k-7576), Biscocha (k-7577).

Varietal type Guacha. Type Papas guachas (k-7557). Tubers white, sometimes with slight anthocyanin on the top, round, highly wrinkled. Eyes deep. Pith yellow. Aerial shoots (haulms) blue-purple. Base of the shoots oval. Apex very small, closed. Scaly leaflet subulate. Pubescence dense with short projecting hairs.

Stem of average length, leaning, slightly colored with anthocyanin. Wings narrow, erect. Many long axillary branches. Leaf of average length, slightly dissected, dark green, not colored with anthocyanin, with lobes touching each other. Terminal lobe small, its shape similar to that of lateral lobes. Base of terminal lobe cordate. Lateral lobes small, oval, dull. Flat leaves found. The first pair of lobes with basiscopic decurrent. Buds fall off.

Amarilla ovalada (k-7566), Importable (k-7540) and Papas guachas (k-7557) are related to this varietal type.

Varietal type La Tayo. Morphologically close to var. *bravipilosum* Buk. et Lechn. Type Amarilla redonda (k-7568). Tubers dark red, round. Eyes of average depth, colored slightly or sometimes somewhat brighter. Brows short. Stolon scar highly depressed. Pith yellow. Aerial shoots (haulms) dark red-purple. Base of the shoots spherical-oval. Apex bent loose, scales leaf-life. Pubescence at the base rare with projecting hairs of average length.

Bush tall. Stem long, erect, moderately colored with anthocyanin. Wings average or slightly undulating. Leaf moderately dissected, with lobes touching each other, colored with anthocyanin on the petiole and in the axis of the stalks of the lobes. Terminal lobe ovate, with cordate base. Lateral lobes oval, with highly acute apex. Stalks of the lobes of average length.

Peduncles long, slightly colored with anthocyanin. Branches of peduncle long, with anthocyanin coloration of the axis of the branches. Upper part of pedicels short, green; lower part long, colored with anthocyanin. Articulation of pedicels colored. Pubescence of pedicels moderate. Calyx colored with anthocyanin, with lanceolate acute tips. Corolla large, red-purple, with white acute tips. Antheral column conical, often deformed. Anthers large, orange. Filaments long, colored with anthocyanin at the base. Style long. Stigma bifid. Ovary slightly colored.

The following are related to this varietal type: k-3381, 6075b, La Tayo (k-7503), Amarilla redonda (k-7568), k-7586, V-833 (sapling of f. *roseum*).

Varietal type Villaroella. Different from var. *villaroella* Buk. et Lechn. Type Yuz. 1980 (sapling of var. *villaroella*). Tubers dark red (beet colored), color beneath the skin, oval. Eyes shallow, colored more intensely. Stolon scar slightly depressed (Fig. 1, e). Pith white or creamish, colored with anthocyanin. Aerial shoots (haulms) dark red-purple. Base of the shoots oval. Apex slightly open. Scaly leaflet phylloid. Pubescence very rare (almost absent).

Bush tall. Stem long, erect highly colored with anthocyanin. Apex of the bush highly colored with anthocyanin. Wings narrow, undulating. Many lateral branches. Leaf of average length, highly dissected, with lobes quite apart from each other. Strong anthocyanin coloration along leaf petiole and at the axix of the lobe stalks. Terminal lobe small, oval. Lateral lobes small, dull, densely pubescent. Stalks of the lobes of average length or long. Leaf petiole of average length.

Flowering profuse. Peduncle long, thin. Inflorescence spreading, branches of peduncle long. Articulation of pedicels in the upper half, colored with anthocyanin. Upper part of pedicels green; lower part colored with anthocyanin. Calyx small slightly colored with anthocyanin densely pubescent. Acute tips of calyx of average size; subulate. Corolla of average size or big, dark red-purple. Acumens brighter (Fig. 3, a). Antheral column conical. Anthers orange. Filaments long. Style long, straight, highly exceeds antheral column in length (up to 4 mm). Stigma bifid. Ovary colored.

Yuz. 1980 (sapling of V_3-297-2), Yuz. 8931 (sapling of V_3-297-3), V-674 are related to this varietal type.

Varietal type Pillan. Type Pillan (k-3466) is a clonal selection from a local variety in Puerto Oktai. Tubers white, round, wrinkled. Eyes deep. Stolon scar slightly depressed. Pith white. Aerial shoots (haulms) dark red-purple. Base of the shoots conical. Apex slightly open. Collar long, thin. Scaly leaflet lanceolate.

Bush tall, well foliated. Stem erect, later leaning, moderately or highly colored with anthocyanin. Many long axillary branches. Internodes of the stem short. Wings narrow. Leaf small, slightly dissected, with lobes apart from each other. Lobes small, first pair of lobes basiscopic decurrent and leans over the terminal lobe. Lobes dull with sharp venation. Stalks of the lobes short. Leaf petiole of average length.

Flowering poor, buds often fall off. Peduncle thin, round, slightly colored with anthocyanin. Pedicels short. Articulation of pedicels in the upper half. Pubescence of pedicels moderate, anthocyanin coloration poor. Calyx large, pubescent with dense oblique hairs. Acute tips of calyx phylloid. Corolla of average size, white, with yellow star. Acute tips of the lobes of corolla small. Antheral column conical. Anthers small (5.5 mm), slightly trapezoid, orange or yellow. Style short, thin, curved. Stigma small, capitulate. Ovary not colored. Berries not formed.

This varietal type is represented by Pillan (k-3466) only.

Varietal type Volcan. Type Volcan No. 52 (k-3475). Tubers round-oval, white. Eyes of average depth. Brows long, not always distinct. Stolon scar highly depressed. Pith pale yellow. Aerial shoots (haulms) red-purple. Base of the shoots spherical-oval. Apex big, slightly open. Scaly leaflet phylloid. Root nodes big. Pubescence moderate with short projecting hairs.

Bush of average height. Stem erect, of average thickness, slightly geniculate, almost without anthocyanin. Wings wide, undulating. Leaf small, slightly dissected, lustrous. Leaf lobes big, similar to lateral lobes. Stalks of the lobes of average length.

Peduncles short. Articulation of pedicels in the upper half. Calyx small, with lanceolate or subulate acute tips. Antheral column regular. Anthers bright orange. Filaments long. Style long, highly exceeds antheral column in length. Stigma large.

Volcan No. 52 (k-3475) and V-454a Morada lisa are related to this varietal type.

Varietal type Quila. Type Quila (k-6061). Tubers white, round, wrinkled. Eyes deep. Stolon scar depressed. Brows not distinct. Pith cream colored. Aerial shoots (haulms) red-purple. Base of the shoots spherical. Apex slightly open. Scaly leaflet lanceolate. Pubescence rare.

Bush tall, slightly or moderately foliated. Stem erect, moderately colored with anthocyanin throughout its length. Wings of average width, undulating. Many axillary branches.

Leaf small, slightly dissected. Lobes apart from each other, rarely touch each other, oval. Terminal lobe small, ovate, with cordate base. Stalks of the lobes short (up to 0.7 cm). Anthocyanin along stalks of upper leaves poor; it is absent in lower leaves. Venation of lobes sharp. Lobes big.

Flowering profuse. Inflorescence big, semi-spreading. Peduncle long, moderately colored with anthocyanin, round. Articulation of pedicels in the upper 1/3, not colored with anthocyanin. Upper part of pedicels green; lower part colored with anthocyanin. Pubescence of pedicels moderate, with short pressed hairs. Calyx small, slightly colored with anthocyanin, moderately pubescent with short pressed hairs. Acute tips of calyx small, subulate.

Corolla big, red-purple. Antheral column conical, anthers orange, small. Style very long (up to 14 mm), of average thickness of thin (up to 8 mm), exceeds antheral column in length. Stigma big, not distinctly bifid.

Quila (k-6061) is related to this varietal type.

Varietal type Magelanes. Morphologically close to the variety Benedicta. Type k-7579. Tubers round-oval, somewhat flat, white but very quickly become green in the light. Eyes of average depth; brows small, not distinct. Stolon scar slightly depressed. Pith pale yellow. Aerial shoot (haulms) red-purple. Base of the shoots conical. Apex small, closed. Scaly leaf lanceolate or phylloid. Root nodes long, thin. Pubescence dense (moderately) with projecting hairs of average height. Long stolons appear quite early.

Bush of average height. Stem erect, afterwards leaning, pure green. Wings narrow, slightly undulating. Nodes of the stem well developed. Axillary branches present. Leaf of average length, slightly dissected, with lobes either touching each other or apart from each other, without anthocyanin coloration. Terminal lobe oval with cordate base. Lateral lobes oval, with highly acute apex. Stalks of the lobes and leaf petiole short. Leaflets round, their shape different from that of the lateral lobes.

Peduncle of average length. Pedicels green, with articulation in the upper half. Anthocyanin absent on the articulation. Pubescence of pedicels dense. Calyx large, green. Corolla small, white. Anthers small, deformed. Filaments long. Style long. Ovary not colored.

The following are related to this varietal type: Talten (k-6093), Blanca de Valdivia (k-6121), Quehuina (k-6126), Chilena (k-6130), Mantequilla (k-6163), Blanca redonda (k-7498), Redonda ovalada (k-7501), Amarilla de Magalanes (k-7507), Amarilla alargada (k-7521), Corahila (k-7535), Bolera (k-7542), k-7579, 7583, 7588, 7598.

Varietal type Huinco (f. *huinco* Lechn.). Type PV-167. Tubers short, napiform, yellow with red-purple anthocyanin at the stolon scar. Eyes deep, slightly colored with anthocyanin. Stolon scar highly depressed. Pith white or creamy. Aerial shoots (haulms) red-purple. Base of the shoots conical. Apex closed. Scaly leaflet lanceolate. Root nodes big.

Bush tall, vigorous. Stem long, thick, erect, later leaning, slightly colored with anthocyanin, which is more intense in the axis of leaf petiole. Wings wide, undulating. Axillary branches many. Leaf big, slightly dissected, with lobes apart from each other. Terminal lobe wide, ovate, with well-developed cordate base. Lobes wide, oval, stalks of the lobes very long (up to 1.7 cm). Leaflets large. In spite of poor dissection there are acroscopic leaflets. Leaf petiole long. Anthocyanin absent on the leaf.

Peduncle long, thick, green. Pedicels very long. Articulation of pedicels in the upper 1/3, colored with anthocyanin. Calyx large, very indistinctly separated from the pedicel. Acute tips of calyx lanceolate or phylloid. Corolla intensely red-purple, large, highly dissected. Bright spots and white margin of lobes present on the acumens. Antheral column cylindrical or slightly conical, compact. Anthers big, thick. Filaments thick. Style very long, highly exceeds antheral column in length (up to 4 mm). Ovary not colored.

PV-169 and PV-167 are related to this varietal type.

Varietal type Pichana. Type Pichana (k-3465). Tubers white, round, with pink tinge at the top. Eyes shallow. Pith white. Aerial shoots (haulms) dark red-purple. Base of the shoots oval. Apex small, slightly open. Scaly leaflet lanceolate. Pubescence very dense, with short projecting hairs.

Bush of average height. Stem leaning, slightly colored with anthocyanin. Wings almost straight. Axillary branches numerous. Leaf moderately or slightly dissected. Lobes oval, touching each other with acute apex. Lobes dark green, dull because of dense pubescence. Leaflets big.

Peduncle short, thin. Inflorescence closed. Articulation of pedicels in the upper half. Pedicel thickens towards calyx. Calyx big, colored with anthocyanin. Acute tips of calyx big, phylloid. Corolla white. Antheral column regular. Anthers small. Berries not formed.

The following are related to this varietal type: k-3403 (b-1), k-3407g (mixture of local varieties from Castro), Bollo de chancho (k-3421), Pichana (k-3465).

Varietal type Palmeta. Type Palmeta (b-908). Identical to variety Up-to-date. Tubers round-oval to oval, white. Eyes shallow or of average depth. Brows of the eyes well developed. Pith white or creamy. Aerial shoots (haulms) red-purple. Base of the shoots spherical-oval to oval. Apex average, open. Scaly leaflet phylloid. Root nodes big. Pubescence moderate with short projecting hairs.

Bush of average height. Stem erect, later leaning, slightly colored with anthocyanin. Wings narrow, straight. Leaf of average size, slightly or moderately dissected with lobes touching each other. Terminal lobe ovate, with cordate base. Stalks of the lobes and leaf petiole of average length. Anthocyanin on leaf absent. First pair of lobes leans over the terminal lobe. Lobes bright-green, dull with sharp venation.

Flowering profuse. Peduncles long, thick, colored with anthocyanin. Inflorescence semi-closed. Articulation of pedicels in the middle not colored. Calyx average sized, moderately colored with anthocyanin, moderately pubescent with hairs of average length. Acute tips of calyx lanceolate. Corolla bright red-purple, with yellow star. Antheral column conical. Anthers big, orange. Filaments thick, of average length. Style thick, straight, with very big bi-trifid stigma.

The following are related to this varietal type: Palmeta (k-345), Leona (k-3385), k-6102, Froma (k-6137), Siete Flores (k-6156), Palmeta (k-7504), Blanca (k-7525) and V-85, 81, 259, 901 and 908.

Varietal type Amarilla redonda. Type Amarilla redonda (k-7573). Tubers white, round-oval to oval. Eyes shallow or of average depth. Pith yellow. Aerial shoots (haulms) red-purple. Base spherical-oval. Apex open. Scaly leaf phylloid. Root nodes big. Pubescence dense.

Bush of average height. Stem leaning, almost without anthocyanin coloration. Wings average. Axillary branches present. Leaf average, slightly dissected, with lobes touching each other, without anthocyanin coloration. Terminal lobe oval (bigger than lateral lobes), oval with cordate base. Lobes big, oval. Stalks of the lobes short. Leaf petiole of average length. Leaflets big.

Peduncle long, thin, green. Flowering moderate. Pedicels green. Articulation of pedicels in the upper half, not colored with anthocyanin. Calyx small, pure green. Acute tips of calyx small, lanceolate or subulate, twisting. Corolla white. Lobes of corolla slightly dissected. Antheral column conical. Anthers large, orange. Style of average length. Stigma capitulate, bifid.

This varietal type is represented only by Amarilla redonda (k-7573).

Varietal type Roja redonda. Type Roja redonda (k-7572). Tubers bright red, round. Eyes of average depth, colored brightly. Brows not always clearly developed. Pith yellow. Base of the shoots conical. Apex loose, closed. Scaly leaf lanceolate. Pubescence rare, with short projecting hairs.

Bush of average height. Stem erect, later slightly leaning, slightly colored with anthocyanin. Wings erect. Axillary branches sometimes present. Leaf large, average or highly dissected, with lobes touching each other. Terminal lobe ovate, with cordate base. Lateral lobes oval, apex of the lobes acute. Stalks of the lobes and leaf petiole of average length. Leaflets big.

Flowering profuse. Peduncles short. Pedicels slightly colored with anthocyanin, moderately pubescent. Articulation of pedicels in the upper half. Calyx average, pubescent with long hairs. Acute tips of calyx small, lanceolate. Corolla red-purple, with bright acumens. Antheral column slightly conical. Style of average length, slightly exceeding antheral column in length. Stigma capitulate. Many berries are formed.

The following are related to this varietal type: k-3339, Morada (k-7533), Rosada ovalada (k-7570), Roja redonda (k-7572).

Varietal type Afrechuda. Type Afrechuda (k-7543). Tubers pink, particularly at the top and in the eyes, oval. Eyes deep. Stolon scar slightly depressed. Pith creamy. Aerial shoots (haulms) red-purple. Base of the shoots conical. Apex open. Scaly leaf phylloid. Pubescence rare, with short projecting hairs.

Stem of average length, thick, leaning, slightly colored with anthocyanin. Wings of average width, straight. Leaf large, moderately dissected, with lobes touching each other. Terminal lobe oval, with cordate base. Lateral lobes oval. Pubescence of lobes dense. First pair of lobes greatly leans over the terminal lobe. Leaflets big. Anthocyanin weak in the axis of the stalks of lobes. Leaf petiole of average length. Flowering for a short duration.

This varietal type is represented by Afrechuda (k-7543) only.

REFERENCES

1. Asseeva, T.V. 1926. *Sortovoi Sostav Krest'yanskikh Posevov Kartofelya v Moskovskoi Gubernii* (Varietal composition of farmers' potato crops in Moscow province), pp. 1–62.
2. Bukasov, S.M. 1933. *Kartofeli Yuzhnoi Ameriki i Ikh Selektsionnoe Ispol'zovanie* (Potatoes of South America and Their Use in Breeding). Leningrad, 153 pages.
3. Bukasov, S.M. 1974. Morfologiya kartofelya (Morphology of the potato). *Trudy Po Prikl. Bot., Gen. i Sel.*, Vol. 53, No. 1, pp. 3–33.
4. Bukasov, S.M. and A.Ya. Kameraz. 1959. *Osnovy Selektsii Kartofelya* (Fundamentals of Potato Breeding). Leningrad, 528 pages.
5. Gern, A. 1961. Sortu Rannyaya sto let (The variety Rannyaya is 100 years old). *Kartofel' i Ovoshchi*, No. 12, pp. 23–25.
6. Zykin, A.G. 1973. Chiliiskii kul'turny kartofel' i ego dikie sorodichi (The Chilean cultivated potato and its wild relatives). *Trudy Po Prikl. Bot., Gen. i Sel.*, Vol. 50, No. 3, pp. 216–240.
7. Kostina, L.I. 1965. Tipy lista sortov kartofelya (Types of leaves of potato varieties). *Trudy Po Prikl. Bot., Gen. i Sel.*, Vol. 37, No. 3, pp. 24–30.
8. Kostina, L.I. 1972. Tipy venchika *Solanum tuberosum* L. (Types of corolla of *Solanum tuberosum* L.). *Trudy Po Prikl. Bot., Gen. i Sel.*, Vol. 48, No. 1, pp. 220–236.
9. Kostina, L.I. and Z.P. Zholudeva. 1971. *Kul'turnaya Flora. Kartofel'* (Cultivated flora. Potato). Leningrad, pp. 305–384.
10. Kostina, L.I. and L.M. Turuleva. 1976. Yuzhnoameri-kanskie formy kartofelya kak iskhodnyi material dlya selektsii (South American forms

of the potato as initial material for breeding). *Byul. VIR*, No. 64, pp. 76–79.

11. Lekhnovich, V.S. 1971. *Kul'turnaya Flora. kartofel* (Cultivated Flora. The Potato). Leningrad, pp. 41–304.
12. Bukasov, S.M., A.G. Zykin, A.Ya. Kameraz, N.F. Bavyko and others. 1976. *Metodicheskie Ukazaniya Po Podderzhaniyu i Izucheniya Mirovoi Kollektsii Kartofelya* (Methodological instructions for the maintenance and study of world collections of the potato). Leningrad, 26 pages.
13. Sinel'nikova, V.N. 1965. Morfologiya lista chiliiskikh form *Solanum tuberosum*) (Morphology of the leaf in Chilean forms of *Solanum tuberosum*). *Trudy Po Prikl. Bot., Gen. i Sel.*, Vol. 37, No. 3, pp. 37–50.
14. Turuleva, L.M. 1975. Khozyaistvennaya kharakeristika chiliiskikh form *Solanum chilotanum* (Economic characteristics of Chilean forms of *Solanum chilotanum*). *Byul. VIR*, No. 47, pp. 52–54.
15. Yuzepchuk, S.V. and S.M. Bukasov. 1929. K voprosu o proiskhozhdeniya kartofelya (The question of the origin of potato). In the book: *Trudy VSes. S"ezda Po Genetike, Selektsii i Semenovodstvu*, Vol. 3, pp. 593–611.
16. Contreras, A.N. 1969. Analisis y pauta de classification de clones de papa recolectados en el sur de Chile. Valdivia, 140 p.
17. Diehl, R. 1938. La pomme de terre. Caracteres et description des varieties. Paris, 157 p.

TRUDY PO PRIKLADNOI BOTANIKE, GENETIKE I SELEKTSII

1978 Vol. 62, No. 1

Geographic Ranges and Local Names of the Primitive Cultivated Potato Species of South America

N.F. Bavyko
Candidate of Agric. Sciences

In the Alpine zone of the South American Andes the diploid species was the first to arise. During the next evolutionary stage, the tetraploid species Solanum andigenum originated. Spontaneous hybridization between the diploid cultivated species and S. andigenum gave rise to the triploid species and later to the only pentaploid species, S. curtilobum. The geographic ranges of the 8 diploid species are described. S. phureja and S. ajanhuiri are endemic to Bolivia. The range of S. stenotomum extends from Bolivia to Peru; the Peruvian species S. goniocalyx occurs in all zones of Peru and is found less frequently in Bolivia; S. canarense is endemic to Ecuador and S. rybinii to Colombia. S. boyacense, S. multijugum and S. kesselbrenneri are distributed from Ecuador to Colombia. The triploid species is distributed from Bolivia to Ecuador. The geographic range of S. curtilobum covers Bolivia, and the southern and central parts of Peru.

For each of the species mentioned the most widely used names are given, in both the local Indian dialects and in Spanish.

The group of primitive cultivated species of South America includes diploid species—*S. phureja* Juz. et Buk., *S. rybinii* Juz. et Buk., *S. kesselbrenneri* Juz. et Buk., *S. canarense* Juz. et Buk., *S. boyacense* Juz. et Buk., *S. multijugum* Buk. et Bavyki, *S. goniocalyx* Juz. et Buk., *S. stenotomum* Juz. et Buk., *S. ajanhuiri*

Juz. et Buk.; triploid species—*S. mammilliferum* Juz. et Buk., *S. tenuifilamentum* Juz. et Buk., *S. chaucha* Juz. et Buk., *S. juzepczukii* Buk. and the pentaploid species *S. curitilobum* Juz. et Buk.

After the work of J. Hawkes [7] in 1944, in which information on the geographic ranges of the cultivated species of the South American potato was given, the works of South American botanists such as C. Ochoa [9, 10, 12], C. Vargas [15], M. Cardenas [5] and D. Untiveras [14] were also published. However, these results were not generalized until recently. Knowledge of the geographic ranges and local names of species helps us to study and establish the phylogeny of species as well as to use them in breeding.

In the present article we tried to ascertain geographic ranges, high altitude zoning and affiliation of local names to individual species.

More detailed information on the collected specimens—their specific determination, local name, altitude of their habitat—is given in the works of Ochoa and Vargas. These works offered an opportunity to verify the boundaries of the geographic ranges of some species; the boundaries of the geographic ranges of *S. phureja*, *S. ajanhuiri*, *S. stenotomum*, *S. juzepczukii* and *S. curtilobum* did not expand much, but the geographic range of *S. goniocalyx* increased considerably.

Among the diploid cultivated species of potatoes two Bolivian species, *S. phureja* and *S. ajanhuiri*, are distinguished by their limited geographic ranges. Only a few isolated specimens were collected beyond the borders of Bolivia—in Peru, in the department Puno which borders Bolivia. These species differ in regard to the vertical zoning of their geographic ranges.

S. phureja Juz. et Buk. (2n=24). An early maturing species without a dormancy period of the tubers. In Bolivia specimens of *S. phureja* were collected by S.V. Yuzepchuk, Hawkes and Cardenas in the department La Paz at an altitude of 1800–2750 m. Isolated specimens of this species were found at an altitude of 3500 m above mean sea level by A.G. Zykin [2] and at an altitude of 3550–3700 m above m.s.l. by Ochoa [10]. The specimens collected by Hawkes in Bolivia at an altitude of 3800 m above m.s.l. were related by him to species *S. phureja* var. *pujeri* Hawk. According to Ochoa [10], isolated specimens of this species were found in Peru (department Puno) at an altitude of 3000–3300 m above mean sea level (Fig. 1). A comparison of this data shows that *S. phureja* could be the rapidly dying, narrow-endemic species in Sorata (Bolivia).

Ochoa found a specimen in Limbanii near Pato carabai on the eastern slopes of the Andes, in the northeast corner of Bolivia, and described it as *S. Limbaniense* Ochoa, the wild originator of *S. phureja*. The tubers are without a dormancy period, on short stolons. This species is distinguished by its very small size, scanty flowering and absence of pubescence. The upper limit of its geographic range is 3700 m above m.s.l. There is no mention of its frost resistance.

Fig. 1. Geographic ranges of the diploid cultivated species of the South American potato.

Ochoa presumed that all cultivated species without a dormancy period of the tubers growing along the eastern slopes of the Cordillera right up to Colombia and Venezuela had a common phylogeny. One of the supposed originators of such a species is *S. amazonicum* Ochoa, which was discovered approximately half-way between Sorata and Colombia.

S.V. Yuzepchuk and S.M. Bukasov [1], J. Hawkes [7], Vargas [15] and Ochoa [9, 10] have done a specific determination of the native varieties and established the conformity of some local names to individual species. Specimens with widely used local names are frequently found in the collections of many authors and they are related by them to one species. R. Mendosa Rendon [4] points out that local names for varieties in the southern and central regions of Peru are often found in the Kechua language. In one of the departments (Cusco) of south Peru the names of varieties are found in a more ancient language, Aimara. In the northern regions of Peru and rarely in the central regions the names of varieties are often found in the Spanish language. Knowledge of these peculiarities makes it possible for us to determine the antiquity of cultivation of the local varieties. Phureja—the local name in the Aimara language—was first found in the work of Cevallos [6] and when translated it means the past. Later on, specimens collected in Bolivia by Yuzepchuk, Hawkes, Ochoa and Zykin under this name with various adjectives, were often related to the species *S. phureja*.

S. ajanhuiri Juz. et Buk. (2n=24). This species has very small width of high altitude zoning. In Bolivia most of the specimens (87%) were collected at an altitude of 3800–4000 meters above m.s.l., and some isolated specimens at a height of 3600–3700 m (Tables 1 and 2; Fig. 1). In Peru only a few specimens of this species were found and they were collected in the department Puno at an altitude of 3850 meters above m.s.l. The number of local names of this species is about 10. The most commonly found names are Ajahuiri, Ckati and Jare. Specimens with these names have been related to species *S. ajanhuiri* by Yuzepchuk, Hawkes and Ochoa. The varietal name Ajahuiri in the Aimara language is widely found in Bolivia and Peru. The other two names—Ckati and Jare—are found very rarely. The antiquity of the geographic range of *S. ajanhuiri* is confirmed by its limited cultivation in Bolivia as well as by the similarity (Ajahuiri) in the varietal names in the Aimara language.

S. juzepczukii Buk. (2n=36). The main geographic range of this triploid species is in Peru and Bolivia (department La Paz). A small number of specimens was collected in the central part of Peru and in the departments Oruro and Kochabamba of Bolivia (Table 1; Fig. 2).

In central Peru specimens were collected at an altitude of 3200–4000 meters above m.s.l., in southern Peru at 3600–4300 meters and in Bolivia at 2800–4150 meters (Table 2). Zykin [2] mentions that specimens of this species were found at an altitude of 4400 meters in Bolivia; most of the specimens (80%) were collected at an altitude of 3600–3900 meters. Among the primitive species of potato, the greatest number of specimens of *S. juzepczukii* (20%) was found at an altitude of 4200–4300 meters above m.s.l.

There are numerous (about 40) local names for this species in comparison with *S. ajanhuiri* and *S. curtilobum*. The most common names for this species are confined to definite countries and departments. For example, in Bolivia (departments La Paz, Kochabamba and Oruro) and in Peru (Puno) this species is widely known by the names Lucki and Kaisalla; in south Peru (Apurimak,

Table 1. Geographic ranges of primitive cultivated species of the potato and the number of local names (according to reports of Ochoa and Vargas)

Country	Department	Number of collected specimens and local names					
		S. ajanhuiri	*S. juzepczukii*	*S. curtilobum*	*S. stenotomum*	*S. goniocalyx*	*S. × chaucha*
Peru	P'yura	—	—	—	—	1 (1)	—
	Libertad	—	—	—	—	8 (3)	17 (12)
Northern zone	Amazones	—	—	—	—	8 (5)	3 (2)
	Kakhamarka	—	—	—	—	8 (4)	14 (2)
	Ankash	—	—	—	—	3 (3)	24 (18)
Central zone	Lima	—	—	1 (1)	—	1 (1)	38 (22)
	Uanuko	—	2 (2)	2 (1)	21 (19)	8 (5)	80 (48)
	Pasko	—	2 (2)	1 (1)	10 (8)	7 (6)	44 (29)
	Khunin	—	3 (2)	3 (1)	24 (15)	28 (16)	132 (81)
	Uankavelika	—	4 (4)	2 (2)	5 (5)	3 (2)	55 (29)
	Ayakucho	—	—	—	10 (9)	8 (4)	11 (11)

Southern zone	Apurimak	—	3 (1)	2 (1)	63 (40)	5 (2)	16 (13)
	Kusko	—	28 (10)	13 (4)	99 (35)	5 (3)	97 (69)
	Puno	5 (3)	49 (27)	14 (7)	67 (38)	3 (3)	42 (42)
	Arekipa	—	—	—	—	—	1 (1)
	Total	5	91	38	299	96	633
Bolivia	La Paz	16 (4)	32 (10)	21 (7)	67 (38)	9 (3)	37 (32)
	Kochabamba	—	2 (2)	3 (3)	1 (1)	—	—
	Oruro	2 (2)	1 (1)	5 (4)	20 (19)	2 (2)	7 (6)
	Potosi	1 (1)	—	1 (1)	—	—	—
	Total	19	35	30	88	11	44
	Grand Total	24 (9)*	126 (40)	68 (20)	387 (150)	107 (30)	677 (350)

Note: Figures in parenthesis indicate the number of local names of the species.

*A number of local names are found both in Peru and Bolivia in some departments of each country and therefore the number of local names is given in common for the species.

Table 2. Spread of primitive cultivated species of the potato according to vertical zoning (according to Ochoa and Vargas)

Altitude zoning, meters above m.s.l.	*S. ajanhuiri*	*S. juzepczukii*	*S. curtilobum*	*S. stenotomum*	*S. goniocalyx*	*S. × chaucha*	Total number of specimens
1600–2200				1	5	6	12
2300					3		3
2600						1	1
2700				1	2	1	4
2800		1		1		14	16
2900						9	9
3000			1	6	6	26	39
3100					3	9	12
3200		1		9	7	13	30
3300				10	5	39	54
3400		1		6	5	58	70
3500				15	11	44	70
3600	1	13	5	90	26	131	266
3700	2	22	11	100	16	56	207
3800	10	40	17	58	6	72	203
3900	6	19	10	32	9	52	128
4000	4	4	2	20		21	51
4100		2	2	10	3	19	36
4200		10		21			31
4300		12	4	7		8	31
Total No. of specimens	23	126	52	387	107	578	1273

Fig. 2. Geographic ranges of cultivated triploid and pentaploid (*S. curtilobum*) species of the South American potato.

Cusco, Puno) it is known by the names Rugui, Ckanchillo and Cuchana; in the central part (Pasko, Khunin, Uankavelika) it goes under the name Shiri.

S. curtilobum Juz. et Buk. (2n = 60). This pentaploid species is often found in southern Peru, rarely in the central part, and in Bolivia (department La Paz).

In the central part of Peru the specimens were collected at an altitude of 3000–3900 meters above m.s.l., in the southern part at 3500–4300 meters and in Bolivia at 3880–4150 meters. The majority of specimens (73%) were collected at an altitude of 3700–3900 meters above m.s.l.

There are not many local names for this species—only about 20. The most common among them are confined to individual zones and departments of a country. In the central part of Peru, from the department Uanuko to Uankavelik, the specimens of this species are found under the name Shiri and in the south under the names Ucumari and Huana, but in Bolivia, Lucki and Choque pito. In spite of the great diversity of local names, some of them are used for two species; therefore it is not always possible to determine the specific affinity of the specimen on the basis of its local name. The local names for *S. juzepczukii* and *S. curtilobum* in particular coincide very often. In some cases names such as Lucki and Shiri are used for these species in the same department, while their specific conformity has been established by many authors. In other cases the name Choque pito is often used for *S. curtilobum* and Rukki (Ruque)—for *S. juzepczukii*.

Among the diploid species one local name for two species is found very rarely. This is noticed in the case of specimens of *S. stenotomum* and *S. goniocalyx* and probably it is just incidental because it is found only in some departments.

S. stenotomum Juz. et Buk. (2n = 24). This diploid species ripens more slowly and gives less yield than *S. andigenum*. *S. stenotomum* is known in numerous forms, most of which (70–80%) were collected in the south (departments Cusco, Puno and Apurimak) and in the central part of Peru (departments Uanuko and Khunin) and also in Bolivia (department La Paz). In the northern part of Peru no specimens of this species were found by any of the researchers. Among the diploid cultivated species, *S. stenotomum* has the widest high altitude zoning—2200–4375 meters above m.s.l.; in the central part of Peru its specimens were collected at an altitude of 2760–4375 m; in the southern part at 2200–4340 m and in Bolivia at 2800–4160 m. The majority of specimens (85%) were collected at an altitude of 3600–4200 m and isolated specimens at an altitude up to 3000 meters above m.s.l.

The differences in ecological conditions resulted in significant diversity in the botanic varieties and forms. Hawkes described nine botanic varieties and 32 forms of *S. stenotomum*. There are a number of local names for this species—about 150; even in some departments of Peru (Cusco, Puno) 35–40 names are used for this species; the most popular names are Pitiquina, Pitiquilla (Piticalla), Chojlla, Phina and Churuspi. It is rare that the same name is found both in Peru and Bolivia; some such names are Chojllo, Surimana and Chiquina. These names are often confined either to one country, or to a department or province. The maximum number of specimens and local names are found in the southern departments of Peru and some of them (Pitiquina, Churuspi) are spread over several departments and provinces but such names are very few. In the central part of Peru one name is rarely found in two departments, and thus we observe great diversity in the specimens and local names for *S. stenotomum*.

Ochoa [11] writes about the continual decrease in the number of diploid species in the central and northern parts of Peru as well as in Bolivia. An especially drastic extinction of a native potato variety has taken place in the case of *S. stenotomum* due to the susceptibility of this species to *Phytophthora* and *Pseudomonas solana. cearum*.

S. goniocalyx Juz. et Buk. (2n=24). The specimens of this species in the collections of S.V. Yuzepchuk and D. Hawkes were obtained from the central part of Peru; only one specimen was obtained from the southern department Apurimak. A comparison of data known to us shows that at present the boundaries of the geographic range of this species have considerably extended into the northern departments of Peru as well as into the south in Bolivia. However, the maximum number of specimens was collected in the central part of Peru at an altitude of 3200–3970 meters above m.s.l. In the northern departments specimens of this species are found at a lower altitude—2230–3600 m, in the southern departments at 3400–3900 m, and in Bolivia at 3590-4100 meters above m.s.l.

Among primitive species, *S. goniocalyx* is closer to *S. stenotomum* regarding the width of high altitude zoning. Most of the specimens of *S. goniocalyx* were collected at an altitude of 3500–3900 meters above m.s.l.; isolated specimens were collected at 2230–2700 m (north Peru) and at 4100 meters above m.s.l. in Bolivia.

S. goniocalyx, cultivated in the hill zone for its high taste quality and yellow pith color of tubers, was obviously brought to the coast at Lima and this is how its local name Limena originated, since it means "from Lima". This could explain the considerable expansion of the geographic range of this species.

There are only about 30 local names for *S. goniocalyx*. Often we find names such as the Kechuanian name Runtu (egg) which corresponds to its pith color, or Spanish names: Amarilla (yellow), Lima, Limena (from Lima) and Zapallo (like the yellow color of the pith of pumpkin). Confinement of these names to certain parts of one country or to certain countries is also observed. For example, Amarilla is widely used in northern and central Peru (it is found in ten departments), Runtu in southern and central Peru (in ten departments) and Lima or Limena in the north. In Bolivia specimens of this species are often found under the name Zapallo. We find these names in the work of Cevallos [6] who was the first to group the varieties of Bolivia according to their economic characters. The much later cultivation of *S. goniocalyx* in the hill zone is suggested by its Spanish name Papa amarilla, i.e. yellow potato, which is the Spanish translation of its Kechuanian name Runtu (egg). This name is mostly found in the initial geographic zone of this species, i.e. in the departments of Lima and Khunin.

Reports on the native varieties in Colombia and Ecuador are not as elaborate as those in Peru and Bolivia. Specimens collected by S.M. Bukasov and D. Hawkes in Colombia and by E. Kesselbrenner and D. Hawkes in Ecuador are very few and are related to species *S. rybinii*, *S. kesselbrenneri*, *S. canarense* and *S. boyacense*. Specimens from Colombia are better represented in the collection of the Colombian Institute (Central Coleccion de la Papa Colombiana). In Colombia the local names, as reported by Montaldo [3], are only 23 and in Ecuador 10. In Colombia we often find the local name Chaucha with adjectives such as colorado, criolla, careta and jardina, while in Ecuador it is Chaucha.

Numerous specimens with the name Chaucha are found in the southwestern departments of Colombia, i.e. on the border with Ecuador, and they are related to the species *S. kesselbrenneri.* In the collection of the VIR, out of 27 specimens with the name Chaucha, 80% are *S. kesselbrenneri.*

Ochoa collected diploid specimens with the name Chaucha in Peru; in the four northern departments, 24 specimens were found at an altitude of 1960–3200 meters above m.s.l., in the central part there were seven specimens at an altitude of 2650–3100 m and in the south there were four specimens at an altitude of 3100–3280 meters above m.s.l. Hence the number of diploid specimens with the name Chaucha becomes drastically lower as we move south. In Bolivia the name Chaucha is reported only in the works of Cevallos [6], Ochoa [11] and S.V. Yuzepchuk. These data show that the name Chaucha is used for certain forms of potato which are mostly found in the north of Ecuador.

In a number of cases the name Chaucha is used with different adjectives, such as Chaucha chuga, Chaucha negra, Chaucha colorada and so on. This may serve as proof of the fact that Chaucha means potato itself, similar to the word "Papa" for the vast territory of the former empire of the Incas.

Triploid species. The specific name *S. chaucha* Juz. et Buk. was first given to the autotriploid species collected by S.V. Yuzepchuk in Peru (department Cusco). A number of forms from other places in Peru and from Bolivia were also related to the species *S. chaucha.* The Peruvian names for these forms are Chimaco, Suittu, Chimo-lomo and Chicchina and the Bolivian names are Sali imilla and Surimana.

A number of triploid specimens with morphological characters clearly distinguishing them from the closer species or hybrids were described as separate species by S.V. Yuzepchuk, S.M. Bukasov and V.S. Lekhnovich—*S. tenuifilamentum* Juz. et Buk., *S. mammilliferum* Juz. et Buk., *S. chocclo* Lechn. and *S. cuencanum* Buk. (Fig. 2). Jackson, Rowe and Hawkes use the specific name *S. chaucha* for all the numerous allotriploids emerging in the territory of Peru as a result of the spontaneous hybridization of *S. andigenum* Juz. et Buk. with the diploid species prevalent there, probably mainly with *S. stenotomum.* Ochoa [12] described triploid specimens of *S.* × *chaucha.* After studying the specimens he grouped them into 30 morphotypes whose phylogeny is quite diverse. The formation of triploids due to the spontaneous hybridization of *S. andigenum* with wild species (for example, *S. bukasovii* or *S. sparsipilum*) is rare. Among the cultivated diploid species involved in the formation of triploids, Ochoa mentions only *S. goniocalyx*, *S. stenotomum* and *S. rybinii*. *S. goniocalyx* gives the triploids an intensive yellow color of the pith and high nutritional qualities. Some of these numerous morphotypes are Huajro and Puca muru rucma, comprising 99–139 specimens which were collected in all zones of Peru at an altitude of 1950–4375 m above m.s.l., but most of the specimens were from an altitude of 3000–4000 m.

Morphotypes Bagazan and Chauchoide, originating from the hybridization and *S. andigenum* with *S. phureja* or with *S. rybinii*, are distinguished by the short dormancy period of their tubers. There are only a few specimens (1–2) and they were collected in the north of Peru (Bagazan) and in the south (Chauchoide) at an altitude of 2800–2950 m. In the morphotypes originating from crosses of *S. andigenum* with *S. stenotomum* the tubers have a comparatively long dormancy period and a white pith.

It is necessary to mention that these morphotypes are distinguished on the basis of the number of specimens, also. Some morphotypes (Arac sapatu, Ayaparara, Ccoe obispo, Ccoyu muru suitu, Chauchoide, Socco unchus, Vitisina and Wuaman uma) are represented by only one specimen each and were collected in different zones of Peru (Ankash-Cusco) at an altitude of 1600–3600 meters above m.s.l. and in Bolivia (department La Paz).

Morphotypes which include between two and nine specimens (Bagazan, Cerrena, Lumbricoide o Huachuco, Paccocha, Pina morado, Pina negro, Pina rojo, Roseolino, Ticla lucma, Yana Rucmaj and Yurac lenle) were collected in different zones of Peru from Libertad to Puno, at an altitude of 2950–4000 meters above m.s.l.

Other morphotypes (Ccoe sullu, Huachuy suitu, Maunoide, Muru puyhuan, Unchuna and Yana lenle) with 12–19 specimens each were collected mostly in southern and central Peru, rarely in all zones of Peru (Huachuy suitu, Muru puyhuan), at an altitude of 2850–4100 m.

A number of morphotypes (Huicti, Puca muru rucma, Puca suitu and Yurac suitu) include 38–99 specimens each, and Huayro 139 specimens; these were collected in all zones of Peru from Libertad to Puno, at an altitude of 1800–4375 meters above m.s.l.

No doubt, among the specimens of *S.* × *chaucha* collected by Ochoa, there can be species already described by the Soviet botanist as well as new forms with unique morphological characters. In this way, according to the data of Ochoa, the maximum number of triploid specimens was collected in the central and southern zones of Peru at an altitude of 3300–3900 meters. Only 9% of the total number of specimens were collected in the northern zone of Peru and 7% in Bolivia. There are about 350 local names for triploid specimens (Table 1). A comparison of the local names of triploid and diploid specimens showed that the names of triploids often coincide with the names of diploid species of *S. stenotomum* and rarely with *S. goniocalyx*.

REFERENCES

1. Bukasov, S.M. 1933. *Kartofeli Yuzhnoi Ameriki i Ikh Selektsionnoe Ispol'zovanie* (Potatoes of South America and their use in breeding). Izd. VIR, Leningrad, pp. 3–146.
2. Zykin, A.G. 1973. Kartofeli Bolivii (Potatoes of Bolivia). *Trudy Po Prikl. Bot., Gen. i Sel.*, Vol. 50, No. 3, pp. 241–267.
3. Lekhnovich, V.S. 1971. Kul'turnye vidy kartofelya (Cultivated species of the potato). In the book: *Kul'turnaya Flora SSSR.* Vol. IX, pp. 41–74.
4. Mendosa Rendon, R. 1974. Geografiya i mestnye nazvaniya aborigennykh sortov Kartofelya Peru (Geography and local names of native potato varieties of Peru). *Trudy Po Prikl. Bot., Gen. i Sel.*, Vol. 53, No. 1, pp. 106–107.
5. Cardenas, M. 1950. Plantas alimenticias nativas de los Andes de Bolivia. Folia Universitaria, No. 2, pp. 2–10.
6. Cevallos, Tovar, Walter. 1914. Noticia Historica y Clacific de papa de Bolivia. Boletin, Vol. X, No. 88.
7. Hawkes, J. 1944. Potato collecting expeditions in Mexico and South America. II Syst. Clasif. of the collections. *Bull. Imp. Bur.*, 142 p.
8. Jackson, M.T., P.R. Rowe and J.G. Hawkes. 1976. The enigma of triploid potatoes. *Am. Potat. J.*, No. 53, p. 395.
9. Ochoa, C. 1955. Expedicion colectora de papas al norte del Peru. *Biota*, Vol. 11, No. 3–4, pp. 1–17.
10. Ochoa, C. 1958. Expedicion colectora de papas cultivadas a la cuenca del lago Titicaca. Progr. de papa. Lima. pp. 1–18.
11. Ochoa, C. 1973. Native potatoes in Bolivia, Chile and Peru. Crop genetic resources in their centers of diversity. FAO.
12. Ochoa, C. 1975. Las papas cultivadas triploides de *Solanum × chaucha* y su distribucion geografica en el Peru. pp. 32–46.
13. Soukup, V. 1939. La classificacion de las papas del Depart. de Puno Minist. de Fomento. Boletin, No. 18, pp. 3–72.
14. Untiveras, D. 1971. Investigaciones en Papa. Universidad National del centro del Peru. Departmento de Publicaciones calle Ferrocarril, No. 469.
15. Vargas, C. 1956. Las papas sud-peruanas. Publ. Univ. Cuzco (Peru), 66 p.

TRUDY PO PRIKLADNOI BOTANIKE, GENETIKE I SELEKTSII

1978 Vol. 62, No. 1

Interspecific Hybridization of the Potato Involving Wild Species of the Series *Acaulia* Juz.

A.Ya. Kameraz
Doctor of Agric. Sciences

M.A. Vavilova, N.A. Zhitlova, V.N. Ivanova
Candidates of Agric. Sciences

Although the wild South American tetraploid species of the series Acaulia are little used in potato breeding, they are of considerable interest in breeding work. This is because they include plants with high frost resistance, plants resistant to some viruses and those which have other economically valuable characters. Hybridization was carried out on a large scale. As a result, germinating seeds were obtained by crossing Acaulia with many species of different ploidy levels, from various series in both South American and North American groups, including the tetraploid S. tuberosum. In the majority of cases a small number of seeds was formed, one per pollinated flower.

When artificially obtained polyploid forms of Acaulia (2n=96) are used for interspecific crosses, better crossability and an increase in the seed number follow. Involving species of Acaulia in breeding work may proceed in different ways. Second crossings of the same with Solanum tuberosum are more promising when aimed at obtaining productive hybrids with a set of economically valuable characters in a comparatively short period of time.

Species of Acaulia as initial material for potato breeding. This series, related by S.M. Bukasov to subsection Acaule, includes the following species: *S. acaule*

Bitt., *S. punae* Juz., *S. depexum* Juz., *S. schreiteri* Buk., *S. uyunense* Card. and *S. aemulans* Bitt. All of them are tall and rosette-shaped, morphologically and physiologically very close to each other (Fig. 1). The northernmost representative—*S. punae* from central Peru—is distinguished by somewhat longer pubescence, the Argentinian *S. depexum* by longer leaves and *S. schreiteri* by much bigger lobes of the leaf.

Fig. 1. *S. acaule* Bitt. (2n=48).

D. Correll [5] suggests that the above-mentioned species are synonyms for *S. acaule*. Some taxonomists consider *S. aemulans*, which has the least dissected leaves, as a subspecies of *S. acaule*. D. Ugent [12] came to the conclusion that tetraploid species of the series Acaulia emerged due to spontaneous hybridization of the diploid species from series *Alticola* Buk. with *S. brevicaule* Bitt. (series *Transaequatorialia* Buk.) and subsequent spontaneous polyploidy.

The species of *Acaulia* inherited a unique type of inflorescence from *Alticola*, but lost any distinct manifestation of the articulation of the pedicel. Abundant berry formation and high frost resistance of the aerial parts of the plants are inherent to all species. Within the limits of the series some specimens resistant to virus X can also be distinguished. Some specimens are relatively resistant to the virus that causes curling of the leaves.

The presence of the above-mentioned characters make the series *Acaulia* particularly valuable for use in potato breeding. Its value increases due to the presence of specimens resistant to potato wart, macrosporiosis, potato scab, oosporosis, black-leg, potato and stem nematodes, as well as due to increased starch content in the tubers. However, species of this series are almost never used in practical breeding work. In the world collection of varieties, we find almost no varieties developed from them. The existing varieties, derivatives of *Acaulia* (for example, Saphir, which is resistant to virus X), do not meet the requirements of commercially valuable varieties as regards their complex of characters.

It follows from the above that in a program of potato breeding work it is essential to pay considerable attention to the use of species of *Acaulia* as initial material. First of all, we must present some results on the crossability of species of this unique and phylogenetically quite distant series with species of other series and then work out ways to increase the effectiveness of hybridization work involving the progeny.

The poor utilization of species of *Acaulia* is mostly related to the extreme difficulty in crossing them with *S. tuberosum*, without which it is not possible to get valuable varieties through interspecific hybridization. Therefore, it is also necessary to investigate intermediate crossings of *Acaulia* with species of different series so that the obtained hybrids can be crossed with *S. tuberosum*. Similarly, it can be presumed that another way, viz. the development of polyploids of the series *Acaulia* artificially and their use in hybridization with *S. tuberosum*, is quite appropriate.

It has already been mentioned that the species of this series are very close. Possibly it would be correct to bring them into one species *S. acaule* as done by Correll [5]. However, since the collection of VIR includes other species of *Acaulia*, the crossing ability of each of them has been examined separately. Most crosses were conducted with *S. acaule* and a few with *S. punae* and others.

The hybridization of Acaulia with wild species of other series. Despite their significant phylogenetic remoteness, tetraploid species of *Acaulia* can be crossed with different South and North American wild species of potatoes of different polyploid levels, belonging not only to different series but also to different subsections (Table 1).

Hybridization with South American species. In the crosses of tetraploid *Acaule* with diploid species of the series *Transaequatorialia* in all cases we got

Table 1. Crossing ability of potato species of the series Acaulia with wild species of other series (1962–1975)

No. of series of species components of Acaulia*	Combination	Number of flowers pollinated	Successful crossings, %	Average number of seeds per berry	Average number of seeds per pollinated flower
1	*S. acaule* × *S. spegazzinii* Bitt.	10	80	154	123
	S. acaule × *S. leptophyes* Bitt.	102	25.2	33.4	6.9
	S. acaule × *S. kurtzianum* Bitt. et Wittm.	162	46.9	21.1	3.5
	S. acaule × *S. famatinae* Bitt. et Wittm.	109	46.5	6.2	3.1
	S. acaule × *S. velascanum* Bitt.	53	22.5	63.2	8.2
	S. acaule × *S. vernei* Bitt. et Wittm.	213	50.7	47.5	25
	S. punae × *S. vernei*	10	30	94	28
2	*S. acaule* × *S. chomatophilum* Bitt.	10	0	—	—
3	*S. acaule* × *S. megistacrolobum* Bitt.	139	45.3	22	9.5
	S. acaule × *S. sanctae rosae* Hawk.	36	27.8	16	4.3
4	*S. acaule* × *S. infundibuliforme* Phil.	13	53.8	75	40
5	*S. acaule* × *S. microdontum* Bitt.	107	42.2	71	28.5
	S. schreiteri × *S. microdontum*	6	33.3	14	4.7
6	*S. acaule* × *S. toralapanum* Card.	12	75	76	57
7	*S. acaule* × *S. tarijense* Hawk.	10	60	148	89
8	*S. acaule* × *S. stoloniferum* Schlechtd.	114	37	42	15
	S. acaule (2n=96) × *S. stoloniferum* (2n=96)	73	85	54	46
	S. acaule × *S. polytrichon* Rydb.	73	66	31	21

	S. acaule (2n=96) × *S. polytrichon* (2n=96)	44	81.8	47	38
	S. punae × *S. stoloniferum*	13	92.3	97	90
	S. acaule × *S. hjertingii* Hawk.	21	9.5	10	0.9
9	*S. acaule* × *S. pinnatisectum* Dun.	181	6.6	14	1
10	*S. acaule* × *S. trifidum* Corr.	76	9.2	1.6	0.14
11	*S. acaule* × *S. cardiophyllum* Lindl.	105	3.8	3.5	0.13
12	*S. acaule* × *S. bulbocastanum* Dun.	126	4.8	1.3	0.06
	S. punae × *S. bulbocastanum*	6	16.7	2	0.33

*South American species. Subsection *Andinum* Buk., series: 1—*Transaequatorialia* Buk.; 2—*Piurana* Hawk.; 3—*Megistracroloba* Card. et Hawk.; 4—*Cuneoalata* Hawk. Subsection *Gigantophyllum* Buk.; 5—series *Simpliciora* Buk. Subsection *Acaule* Buk.; 6—series *Alticola* Buk. Subsection *Tarijense* Buk.; 7—series *Tarijensa* Corr. North American species. Subsection *Longipedicellatum* Buk.; 8—series *Longipedicellata* Buk. Subsection *Pinnatisectum* Buk.; 9—series *Pinnatisecta* (Rydb.) Buk. Subsection *Trifidum* Buk.; 10—series *Trifida* Corr. Subsection *Cardiophyllum* Buk.; 11—series *Cardiophylla* Buk. Subsection *Bulbocastanum* Buk.; 12—series *Bulbocastana* Rydb.

berries with seeds. The number of successful crosses varied from 22.5 to 80%, and the average number of seeds per berry varied from 21.1 to 154 while that of each pollinated flower varied from 3.1 to 123. The combination *S. acaule* × *S. spegazzinii* is distinguished by a high percentage of successful crossings and number of seeds per berry and per pollinated flower. However, in this combination poor germination of seeds (13%) was observed. In some combinations, for example, *S. acaule* × *S. famatinae* and *S. acaule* × *S. kurtzianum*, the berries had very few seeds. Therefore, despite a relatively high percentage of successful crosses (46.5 and 46.9) the average number of seeds per pollinated flower was low (3.1 and 3.5).

Of considerable interest are the crosses of *Acaulia* with *S. vernei*, which is distinguished for its resistance to Phytophthora under field conditions. In the combination *S. acaule* × *S. vernei*, successful crosses were 50.7%, and the average number of seeds per berry and per pollinated flower was 47.5 and 25, respectively, with their germination rate being 30% (Fig. 2, a).

Diploid *S. microdontum* from the series *Simpliciora* was taken for hybridization with *Acaulia*. In the combination *S. acaule* × *S. microdontum* there were 42.2% successful crosses; the average number of seeds per berry and per pollinated flower was 71 and 28.5, respectively; germination of seeds was 51.6%, with reduced germinative vigor (on 25–30th day after sowing).

All hybrid plants examined cytologically by M.A. Vavilova and N.A. Zhitlova [2], as was expected, were triploids (2n = 36) and sterile. They bloomed abundantly but did not form berries through self-pollination. Pollen, as a rule, did not change color with acetocarmine. In some cases an extremely insignificant number of colored pollen grains was observed. According to Swaminathan [10, 11], in triploid hybrids of crosses of *S. acaule* × *S. microdontum* there are 0–2% fertile pollen grains. At the same time, amphidiploid plants of this combination are fertile: they have, on an average, 61% fertile pollen grains.

Crosses of *S. acaule* with other wild diploid species of different series in the subsection *Andinum* show that hybrid berries were obtained except in the case of crossing with *S. chomatophilum* (with a small number of pollinated flowers. Table 1). The involvement of *S. infundibuliforme* (series *Cuneoalata* Hawk.) in hybridization gave more positive results. In this case the number of successful crosses was 53.8%; the average number of seeds per berry was 75 and per pollinated flower 40; germination of seeds was 95%. *S. megistacrolobum* and *S. sanctae rosae*, as components of crossing, gave rather poor results. In combinations involving them the above-mentioned indices were 45.3 and 27.8%; 22 and 16; 9.5 and 4.3; 44 and 24%, respectively. Hybrids of these crosses were also triploids which were sterile and the seeds generally had a low germinative vigor. For example, seeds of *S. acaule* × *S. sanctae rosae* and *S. acaule* × *S. toralapanum* began to germinate on the 40–45th day after sowing.

Fig. 2. Plants of interspecific hybrids involving *S. acaule* Bitt.:

a—F_1 of *S. acaule* (2n=48) × *S. vernei* Bitt. et Wittm. (2n=24);
b—F_1 of *S. acaule* (2n=48) × *S. stoloniferum* Schlechtd. (2n=48);
c—F_1 of *S. acaule* (2n=48) × *S. bulbocastanum* Dun. (2n=24);
d—F_1 of *S. acaule* (2n=96) × *S. tuberosum* L. (2n=48).

Hybrid plants obtained by crossing *Acaulia* with different species of the subsection *Andinum* have an intermediate habit of their bush with a height of 20–25 cm. Hybrids of *S. acaule* × *S. megistacrolobum* retained the well manifested rosette-like shape of the bush. In contrast to the female component of the cross, the hybrids had distinct peduncles measuring 0.5–5.0 cm in length. The flowers were bigger than those of *S. acaule.*

Among species of other South American subsections, the diploid species *S. chacoense* Bitt. (series *Glabrescentia* Buk.) and *S. tarijense* Hawk. (series *Tarijensa* Corr.) were involved in crossings with *Acaulia* with positive results. When a small number of flowers of tetraploid *S. acaule* were pollinated with pollen of the diploid *S. tarijense*, we got 60% successful crosses, and 148 and 89 seeds per berry and per pollinated flower, respectively, but with reduced germination of seeds—17.8%.

Species of *Acaulia* were similarly crossed with wild North American (mostly Mexican) species. Of considerable interest was the hybridization with the Phytophthora-resistant species of series *Demissa* and *Longipedicellata.* Even in the pre-World War years we and other researchers crossed *Acaulia* with hexaploid *S. demissum.* In such crosses berries were formed with viable seeds. Propach [9], reporting the difficulty in involving *S. acaule* (2n = 48) in hybridization, mentioned that this species is relatively easily crossed with the diploid species *S. chacoense*, *S. verrucosum* and *S. phureja*, with the tetraploids *S. ajuscoense* and *S. fendleri* and with the hexaploid *S. demissum.* In the present article we will outline the data which we have obtained since 1962. Therefore, we have not included material on the hybridization of *S. acaule* with *S. demissum.*

Hybridization of *Acaulia* with some species of the series *Longipedicellata* was quite successful in all cases (Table 1). Poor compatibility was noticed in the combination *S. acaule* × *S. hjertingii:* successful crosses 9.5%, average number of seeds per berry 10 and per pollinated flower—0.9. In the hybridization of the tetraploid *Acaulia* with the tetraploids *S. stoloniferum* and *S. polytrichon* these indices varied within limits of 37–92.3%, 31–97 and 15–90, respectively, while the germination of seeds was 48.5–65.7% (Fig. 2, b).

Hybridization between the polyploid form of *S. acaule* (2n = 96) and the polyploids *S. stoloniferum* (2n = 96) and *S. polytrichon* (2n = 96) gave a much higher percentage of successful crosses and more seeds per berry and per pollinated flower.

The formation of nonreducing gametes is obviously inherent in the tetraploid *Acaulia.* Therefore, in hybridization with the tetraploid *Longipedicellata*, hexaploid hybrids are often developed.

Similarly hybridization of the tetraploid *Acaulia* with diploid species of the North American group of series *Pinnatisecta. Cardiophylla*, *Trifida* and *Bulbocastana* (Fig. 2, c), which were related to different subsections by S.M. Bukasov, was quite successful. Here, in all cases we observed a very small percentage of successful crosses—3.8–9.2% (only in the combination *S. punae* × *S. bulbocastanum* was it 16.7%, but with a very small number of pollinated flowers), a very low average number of seeds per berry—1.3–3.5 (only in the combination *S. acaule* × *S. pinnatisectum* was it 14 with germination of seeds 42.6%) and per pollinated flower—0.06–1.0 (see Table 1).

Hybrid progeny in tetraploid-diploid crosses (4x × 2x) are mostly triploid; tetraploids are seen rarely. For example, out of 10 plants studied from the combination *S. acaule* × *S. bulbocastanum*, seven were triploids and three tetraploids, whereas in the combination *S. acaule* × . *S. pinnatisectum* out of six plants four were triploids and two tetraploids [2].

Analyzing 72 hybrids of *S. acaule* × *S. bulbocastanum*, Hermsen [6] found 59 triploids and 13 tetraploids. As mentioned above, generally only a triploid progeny is formed from crosses of *Acaulia* with the diploid species of other series. A sound conclusion can be made here: in crosses with diploid species, *S. acaule* often behaves as a physiologically functional diploid. However, even in such combinations the emergence of some tetraploid plants is noticed.

Instances are known, when in crosses of tetraploids (*S. punae* × *Smyslovskii*) with diploids (*S. rybinii* × *S. bukasovii*) there emerged hexaploids which were obviously the result of the presence of nonreducing gametes in both the components. While the plants of species of the series *Acaulia* and *Longipedicellata* are self-fertile and produce a large number of berries through self-pollination, the hybrid plants obtained as a result of crossing *Acaulia* with diploids *S. tarijense*, *S. microdontum*, *S. vernei*, *S. pinnatisectum* and *S. bulbocastanum*, as well as with tetraploid species of the series *Longipedicellata* are found to be sterile.

The hybridization of Acaulia with S. tuberosum L. Tetraploid species of *Acaulia* are very difficult to cross with the tetraploid *S. tuberosum*. In order to improve crossability, various special means are adopted such as advance polyploidization of the wild species.

Barring a few rare exceptions, tetraploid species of *Acaulia*, when used as the male parent in the cross, do not give positive results. According to L.A. Dremlyug [3], this is to a great extent associated with the presence of very short pollen tubes; these pollen tubes extend up to the embryonal sac only in a few isolated cases.

Our data (1967–1975) on the crossability of octoploid and tetraploid *S. acaule* with *S. tuberosum* (2n = 48) shows that in the first case the number of successful crosses increases twofold (39 and 19%). The average number of seeds per berry (15.2 and 9) and per pollinated flower (5.9 and 1.7) also increases, although it is still not high (Fig. 2, d).

Cytological analysis of a small number of plants from the crosses of tetraploid *S. acaule* with tetraploid *S. tuberosum* revealed the presence of exclusively tetraploid hybrids, which formed numerous berries. In such cases 70–89% of the pollen grains were colored.

Among hybrids from crosses with the polyploid form of *S. acaule* (2n=96), we observed segregation in the chromosome number from 48 to 72 (also with plants which had 56 and 60 chromosomes). In all cases fruit formation was good. The lowest quantity of colored pollen grains (37–60%) was in the 72 chromosome hybrids [2].

In crosses of tetraploid *Acaulia* with *S. tuberosum*, Propach [9] observed many nonreducing ovicells in *Acaulia* with 48 chromosomes. As a result, instead of the expected 48 chromosomes, the hybrids had 72 chromosomes and were often fertile.

The hybrid developed by M.S. Blagovidova in the '30s from the cross *S. punae* (2n=48) × *S. tuberosum* (Smyslovskii) was tetraploid. It gave only a few berries through self-pollination. However, in subsequent hybridization with the variety Centifolia the progeny had 72 chromosomes. Obviously, the progeny was obtained as a result of the fertilization of nonreducing gametes of the hybrids F_1 of *S. punae* × *Smyslovskii*. Hexaploid progeny of hybrids mostly inherit the characters of the wild species. Subsequent crossings with *S. tuberosum* lead to a reduction in the chromosome number up to pentaploid and tetraploid levels.

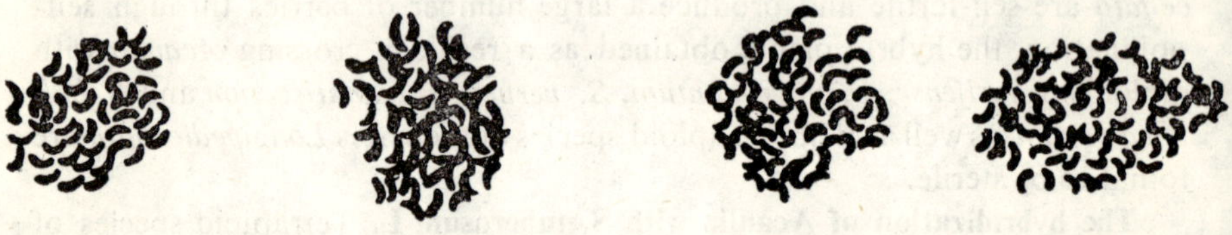

2n=48 2n=56 2n=60 2n=72

Fig. 3. The chromosome numbers of interspecific hybrids of potato from the crosses of *S. acaule* (2n=96) with different varieties of *S. tucerosum*.

An improvement in the crossability of *S. acaule* (2n=96) with *S. tuberosum* (2n=48) is reported by many authors, while back crossing is not successful (8, 11 and others). According to the data of M. Swaminathan [11], in the combination *S. acaule* (2n=96) × *S. tuberosum* (2n=48) there were 37.2% successful crosses while the average number of seeds in the berry was 31. Irikura [7] did not get berries from the straight and even reciprocal crosses of *S. acaule* (2n=48) with *S. tuberosum* (2n=48). In the combination *S. acaule* (2n=96) × *S. tuberosum* berry-formation was 67%; the average number of seeds per berry was 48.8 and per pollinated flower 32.5. Different chromosome numbers were observed in the hybrids (Fig. 3).

The above-mentioned data testify to the possibility of using tetraploid as well as polyploid (octoploid) forms of *S. acaule* in crosses with *S. tuberosum*.

According to the data of some researchers, in the latter case the success of hybridization increases considerably.

Repeated hybridization of F_1 (Acaulia × S. tuberosum) with different varieties of S. tuberosum. During the process of obtaining backcrosses and concurrent backcrosses of *Acaulia* with the involvement of varieties of *S. tuberosum* only, the crossability increases considerably (Table 2). This is observed in both cases with the involvement of both tetraploid and polyploid in the primary crossing. In combinations like (*S. acaule* 96 × *S. tuberosum*) × *S. tuberosum* and (*S. punae* 96 × *S. tuberosum*) × *S. tuberosum* the successful crosses were 81–65%, the average number of seeds per berry was 60–150 and per pollinated flower—49–97. Good results were obtained in this combination: *S. tuberosum* × (*S. punae* 96 × *S. tuberosum*). It was only with the involvement of tetraploid *S. aemulans* (with a small quantity of pollinated flowers) that the results of the last crosses became considerably worse and in getting the backcrosses the average number of seeds per berry was 6.4–17, and per pollinated flower—3.8–9.3.

Table 2. Results of concurrent crossings of F_1 (Acaulia × S. tuberosum) with different varieties of S. tuberosum (2n=48), 1962–1975

Combination	Number of pollinated flowers	Successful crosses, %	Average number of seeds per	
			berry	pollinated flower
(*S. acaule* 96 × t) × t	224	81	60	49
[t × (*S. acaule* 48 × t)] × t	116	45	98	44
(*S. punae* 96 × t) × t	23	65	150	97
t × (*S. punae* 96 × t)	84	69	34	23
[t × (*S. punae* 96 × t)] × t	270	27	12	3.1
(*S. aemulans* 48 × t) × t	5	60	6.4	3.8
t × (*S. aemulans* 48 × t)	9	56	17	9.3

Note: t—*S. tuberosum* (2n=48); 48, 96 chromosome number (2n) of the components of crosses.

Recurrent backcrosses are affected to a great extent by the peculiarities of the components of the crosses, viz. the backcross and *S. tuberosum*, particularly in the last cross. According to our data, while developing recurrent backcrosses [*S. tuberosum* × (*S. acaule* 48 × *S. tuberosum*)] × *S. tuberosum* the crossability can sometimes be better than in making the recurrent backcrosses [*S. tuberosum* × (*S. punae* 96 × *S. tuberosum*)] × *S. tuberosum* (see Table 2). Naturally, the data presented in the table do not characterize with necessary reliability the crossability in combinations of different types, because in the above variants the quantity of pollinated flowers was different, different varieties were involved in the crosses, etc. However, the plant breeder can draw one conclusion with complete certainty: in the case of recurrent crossing of hybrids of *Acaulia* with varieties of *S. tuberosum*, the percentage of success increases considerably as do the average number of seeds per berry and per pollinated flower.

In this way, in the case of the species of *Acaulia*, the breeding process can be structured on the basis of recurrent crossings with *S. tuberosum* without involvement of any other varieties. Seeds for further breeding work can be obtained at every succeeding stage of hybridization.

The crossing of hybrids (F_1 Acaulia × wild species) with S. tuberosum. The above-mentioned data obtained by crossing *Acaulia* species with various wild potato species of other series (see Table 1) can be of practical use only if they are involved in further hybridization with *S. tuberosum*. Similar crosses were made by us (Table 3).

In certain cases allopolyploids were developed with the help of colchicine in order to improve crossability. Hexaploid hybrids obtained from triploids are examples of such allopolyploids: *S. acaule* × *S. spegazzinii*, *S. acaule* × *S. vernei*, *S. acaule* × *S. infundibuliforme*, *S. acaule* × *S. megistacrolobum*, *S. acaule* × *S. microdontum*. Some of these hexaploid hybrids developed berries through self-pollination.

The crossability of F_1 (*Acaulia* × wild species) with different varieties of *S. tuberosum* is not high, particularly when components of *Acaulia* in the primary cross were the South American species of the series *Transaequatorialia*, *Simpliciora*, *Cuneoalata*, *Alticola* and *Tarijensa* (Table 3). Crossings in a number of combinations gave no results. In the formation of the berry, the quantity of seeds per pollinated flower was insignificant.

Slightly better results were obtained in the following combinations: (*S. acaule* × *S. kurtzianum*) × *S. tuberosum* (successful crosses 68%, the average number of seeds per berry was 28 and per pollinated flower 19); *S. tuberosum* × (*S. acaule* × *S. spegazzinii*) 2n = 72 (25–7%; 71 and 18.3); *S. tuberosum* × (*S. acaule* × *S. microdontum*) 2n = 72 (42.8%; 11 and 4.8); *S. tuberosum* × (*S. acaule* × *S. infundibuliforme*) 2n = 72 (20%; 25 and 5.1); *S. tuberosum* × (*S. acaule* × *S. megistacrolobum*) 2n = 72 (33.9%; 61 and 21). Generally, doubling the chromosome number of the triploid hybrid improves its crossability with tetraploid *S. tuberosum*.

Hybrids from crosses of tetraploid *Acaulia* with the tetraploid species of *Longipedicellata* (*S. stoloniferum* and *S. polytrichon*) in further crossings with the tetraploid *S. tuberosum* give berries with seeds in which the average number of seeds per pollinated flower is extremely low.

The success of hybridization increased quite significantly when F_1 hybrids between the polyploid species *Acaulia* and *Longipedicellata* were used for succeeding crosses with *S. tuberosum*. For example, in the combinations (*S. acaule* 96 × *S. stoloniferum* 96) × *S. tuberosum* and (*S. acaule* 96 × *S. polytri-*

Table 3. The crossability of interspecific hybrids of the potato F_1 (Acaulia × wild species) with S. tuberosum and other cultivated species (1964–1975)

Combination	Number of pollinated flowers	Successful crossings, %	Average number of seeds per berry	Average number of seeds per pollinated flower
(*S. acaule* × *S. kurtzianum*) × t	104	68	28	19
(*S. acaule* × *S. spegazzinii*) 36 × t	13	0	—	—
(*S. acaule* × *S. spegazzinii*) 72 × t	31	3.2	30	1
t × (*S. acaule* × *S. spegazzinii*) 72	132	25.7	71	18.3
t 24 × (*S. acaule* × *S. spegazzinii*) 72	9	0	—	—
(*S. acaule* × *S. leptophyes*) × t	37	21.6	16	3.6
(*S. acaule* × *S. vernei*) 36 × t	53	0	—	—
(*S. acaule* × *S. vernei*) 72 × t	5	0	—	—
t × (*S. acaule* × *S. vernei*) 72	70	11.4	14	1.6
(*S. punae* × *S. vernei*) × t	148	0.7	1	0.006
(*S. acaule* × *S. famatinae*) × t	92	68	21	14
(*S. acaule* × *S. microdontum*) × t	222	0	—	—
(*S. acaule* × *S. microdontum*) 72 × t	82	12.2	22	2.7
t × (*S. acaule* × *S. microdontum*) 72	14	42.8	11	4.8
t 24 × (*S. acaule* × *S. microdontum*) 72	101	10.9	6	0.7
(*S. acaule* × *S. microdontum*) 72 × *S. rybinii*	37	10.8	1.5	0.16
S. rybinii × (*S. acaule* × *S. microdontum*) 72	33	3	2	0.06
(*S. acaule* × *S. infundibuliforme*) 72 × t	32	3.1	12	0.38
t × (*S. acaule* × *S. infundibuliforme*) 72	15	20	25	5.1
t × (*S. acaule* × *S. megistacrolobum*) 72	68	33.9	61	21
t 24 × (*S. acaule* × *S. megistacrolobum*) 72	42	45.2	5	2.4
(*S. acaule* × *S. tarijense*) × t	122	4.1	1.2	0.05
(*S. acaule* × *S. stoloniferum*) × t	2259	22	6.2	1.37
(*S. acaule* 96 × *S. stoloniferum* 96) × t	54	82	32	26
(*S. punae* × *S. stoloniferum*) × t	1434	9	1.7	0.15
(*S. acaule* × *S. polytrichon*) × t	606	35.3	12	4.2
(*S. acaule* 96 × *S. polytrichon* 96) × t	40	82.5	39	32
(*S. acaule* × *S. pinnatisectum*) × *S. phureja*	6	100	13.7	14
(*S. acaule* × *S. pinnatisectum*) × *S. andigenum*	7	0	—	—
(*S. acaule* × *S. pinnatisectum*) × t*	190	0	—	—
(*S. acaule* × *S. cardiophyllum*) × t*	98	0	—	—
(*S. acaule* × *S. bulbocastanum*) × t*	137	0	—	—

Note: Here and afterwards 24, 36, 72, 96—chromosome number (2n) of the components of crosses.

*Only parthenocarpic berries were obtained in the cross.

chon 96) × *S. tuberosum* the percentage of successful crosses was 82%; the number of seeds per berry and per pollinated flower was 32–39 and 26–32, respectively.

In the crosses of the tetraploid *S. tuberosum* with the first generation hybrids from the tetraploid *S. acaule* and diploid *S. pinnatisectum*, *S. cardiophyllum* and *S. bulbocastanum* all the berries that formed were without seeds. In order to improve the crossability of these species it is necessary to resort to some special measures of overcoming incompatibility, such as advance polyploidization.

In the combination (*S. acaule* × *S. pinnatisectum*) *S. phureja* 24 (with a very small quantity of pollinated flowers) we got positive results, whereas use of the tetraploid *S. andigenum* in subsequent crossings to get a backcross gave a negative result.

Double and triple crossing of S. tuberosum with hybrids (F_1 Acaulia × wild species). Recurrent crossings of the hybrids of *Acaulia* with *S. tuberosum* increase crossability considerably (Table 4). The success of crossing in recurrent hybridization with *S. tuberosum* is largely affected by the component

Table 4. Recurrent crossings of interspecific hybrids of F_1 potato (Acaulia × wild species) with S. tuberosum (2n=48), 1965–1975

Combination	Number of pollinated flowers	Successful crossings, %	Average number of seeds per berry	Average number of seeds per pollinated flower
[t × (*S. acaule* × *S. spegazzinii*)72] × t	30	20	32	6.4
[t × (*S. acaule* × *S. spegazzinii*)] × *S. rybinii*	35	5.7	1	0.06
[t × (*S. acaule* × *S. microdontum*)] × t	52	0	—	—
[t × (*S. acaule* × *S. infundibuliforme*) 72] × t	16	6.3	15	0.94
[(*S. acaule* × *S. stoloniferum*) × t] × t	29	0	—	—
[(*S. acaule* 96 × *S. stoloniferum* 96) × t] × t	82	36.6	32.5	11.8
t × [(*S. acaule* × *S. stoloniferum*) × t]	269	49	36	18
{[(*S. acaule* × *S. stoloniferum*) × t] × t} × t	42	36	26	9
{t × [(*S. acaule* × *S. stoloniferum*) × t]} × t	951	32	27	8.7
t × {t × [(*S. acaule* × *S. stoloniferum*) × t]}	435	41	24	9.8
[{t × [(*S. acaule* × *S. stoloniferum*) × t]} × t] × t	2	100	28	28
t × [{t × [(*S. acaule* × *S. stoloniferum*) × t]} × t	27	48	34	17
[(*S. acaule* 96 × *S. polytrichon* 96) × t] × t	69	30.4	46.1	14
[(*S. punae* × *S. stoloniferum*) × t] × t	103	53	106	56
t × [(*S. punae* × *S. stoloniferum*) × t]	217	59	48	28
[(*S. punae* × *S. stoloniferum*) × t] × t	9	78	50	39
{t × [(*S. punae* × *S. stoloniferum*) × t]} × t	472	34	18	6
t × {t × [(*S. punae* × *S. stoloniferum*) × t]}	245	35	13	4.5
t × {[(*S. punae* × *S. stoloniferum*) × t] × t}	25	56	7	4

(wild species) of the primary crossing with *Acaulia*, as well as by the peculiarities of the components at all stages of the succeeding crossings.

The involvement of some South American species in hybridization with *Acaulia* in single (Table 3), as well as concurrent crossings (Table 4), with *S. tuberosum* gives less favorable results than the involvement of species of *Longipedicellata* of the North American group in the primary crossing. We were determined to involve hybrids of *Acaulia* × *Longipedicellata* in crosses with *S. tuberosum*. These hybrids were obtained between specimens of *Acaulia*, which were preliminarily singled out on the basis of their resistance to virus X, and specimens of *Longipedicellata*, which were distinguished for their resistance to viruses Y and A. Hybrids between the above-mentioned series of species, which had a complex resistance to these three viruses (X, Y, A), were involved in crosses with *S. tuberosum*.

Fig. 4. A plant of the hybrid *S. tuberosum* × [(*S. acaule* 48 × *S. stoloniferum* 48) × pollen mixture of *S. tuberosum*].

In almost all cases, hybrids of *Acaulia* × *Longipedicellata* in double-triple crossings with *S. tuberosum* developed berries with seeds. The average number of seeds per berry varied from 7 to 106 and per pollinated flower from 0.2–4 to 56 (see Table 4; Fig. 4).

In this way, despite the distinctive nature, phylogenetic remoteness and, as a rule, difficult crossability of any wild species of the potato of series *Acaulia*, they can be involved in hybridization with diverse wild species of the potato as well as directly with *S. tuberosum*. These primary hybrids can be successfully used in further crossings with *S. tuberosum*. All these facts indicate that at present there is hardly any difficulty in using species of *Acaulia* in breeding work.

SOME AGRO-ECONOMIC PROPERTIES OF HYBRIDS OF ACAULIA

Yield and starch content. The first generation of hybrids of tetraploid *Acaulia* from crosses with many other wild species under long-day conditions generally does not develop tubers or forms 1–2 very small tubers (Table 5). When cultivated in flower pots under an artificially reduced length of day, the yield from one plant was 60–115 g. The starch content varies significantly. It depends to a considerable extent not only on the initial specimen of *Acaulia*, but also on the component of crossing, i.e. the variety of *S. tuberosum*. Stolons are very long and spreading.

Table 5. Frost resistance and yield of hybrids from crosses with S. acaule (2n=48), 1969–1975

Combination	Frost resistance in points: −3°×4 hours	Frost resistance in points: −4°×4 hours	Yield from one plant, g
ac×t	9	9	0–100
ac×t×t	5–7	3–7	300–1500
ac×t×t×t	3–7	3–7	400–2000
ac×*S. toralapanum*	9	9	—
ac×*S. vernei*	9	7–9	—
ac×*S. megistacrolobum*	9	7–9	—
ac×*S. microdontum*	9	5–9	—
(ac×*S. microdontum*)×t	5–7	3–7	50–200
t×(ac×*S. microdontum*)	7	3	50–300
(ac×*S. microdontum*)×t×t	5	3	100–400
ac×*S. spegazzinii*	7–9	7	—
t×(ac×*S. spegazzinii*)	7	5–7	50–300
ac×*S. infundibuliforme*	5–7	5–7	—
(ac×*S. infundibuliforme*)×t	3–5	3–5	50–200

Note: Here and afterwards ac—*S. acaule*.

One crossing of F_1 (*Acaulia*×wild species) with *S. tuberosum* increases the yield, but it does not exceed 200–400 g per plant (Table 5). Two or three

Table 6. Some agro-economic characters of the first tuber reproduced from hybrids from concurrent crossings of F_1 Acaulia (2n=48) × S. stoloniferum with S. tuberosum (2n=48), 1968–1970

Combination	Number of hybrids	Yield from one plant, g		Starch content, %	
		average	variation	average	variation
B_1 [(ac × st) × t] × t	14	235	50–550	15.9	13–19
B_1 t × [(ac × st) × t*]	23	341	50–765	16.5	13–21
B_2 t × [(ac × st) × t*]	8	251	80–600	16	10.2–20.6
B_1 {t × [(ac × st) × t]} × t	68	224	50–1000	13.8	10–19
B_1 t × {t × [(ac × st) × t*]}	15	247	50–600	14.1	10.5–17
B_1 [(*S. punae* × st) × t*] × t	8	445	50–880	13.5	11–15.4
B_1 t × (*S. punae* × st) × t*	2	460	320–600	13.6	12.9–14.4
B_1 {t × [(*S. punae* × st) × t*]} × t	8	286	90–750	16.3	12.9–23

Note: Here and afterwards st—*S. stoloniferum* (2n=48).
*Pollen mixture of improved varieties of *S. tuberosum.*

crossings with *S. tuberosum* enable us to increase the yield of hybrids to an even greater extent, but nevertheless not up to the required level.

Table 6 shows yield data on the first tuber reproduction of some hybrids from double-triple crossings of *Acaulia* × *S. stoloniferum* with *S. tuberosum*. These hybrids were better in corresponding combinations when grown under field conditions. In tuber reproduction their maximum yield per plant was only between 550–1000 g, while the minimum yield was 50 g in almost all combinations.

The starch content of tubers in various combinations also varied significantly. In some hybrids it was as high as 20–23%.

Yield as well as starch content of the tubers are to a great extent affected by the components of crossing, i.e. the varieties of *S. tuberosum*. Therefore, some hybrids from double crossings yield more than those from a triple crossing with *S. tuberosum*.

Concurrent hybridization with *S. tuberosum* reduces the length of the stolons, which nevertheless remains considerable and gives the bush a less compact shape. In order to get hybrids with the desired complex of agro-economically valuable characters, further crossings are essential with specially selected early varieties of *S. tuberosum*, which should be appropriately used as the female parent.

While developing bispecific hybrids from crosses of *Acaulia* with *S. tuberosum* it is possible to build the required complex of agro-economically useful characters in a shorter period of time than in the case of trispecific hybrids, when hybridization with *S. tuberosum* is preceded by crossing between two wild species.

In the F_1 of *Acaulia* (2n = 48) × *S. tuberosum* (2n = 48) the yield is very low under long-day conditions; it varies within a range of 0–100 g of small tubers per plant. However, in the backcrosses (*S. acaule* × *S. tuberosum* × *S. tuberosum*) it is up to 1500 g in some hybrids while in concurrent backcrosses (*S. acaule* × *S. tuberosum* × *S. tuberosum* × *S. tuberosum*) it is as high as 2000 g per plant (see Table 5). Starch content varies within a wide range—from 10.7 to 20% in backcrosses and from 13.2 to 23% in recurrent backcrosses. The stolons are less spreading than in trispecific hybrids. However, even in recurrent backcrosses hybrids with compact stolons are rarely found.

The yield of the hybrids in F_1 greatly increases if the polyploid (2n = 96) form of *Acaulia* is used for crossing with *S. tuberosum* (Table 7). In the F_1 of crosses of *S. acaule* (2n = 96) with different varieties of *S. tuberosum* the yield of some hybrids in almost every combination is as high as 1000 g in the first tubers produced and sometimes it is even 1200 g. The minimum yield is 10–50 g, while the average yield within each combination varies from 75 to 150 g. The starch content varies from 10 to 18.5%.

In backcrosses the maximum yield of hybrids in different combinations reached 1200–1500 g, i.e. approximately up to the same levels as in backcrosses

Table 7. Some agro-economic characters of the first tuber produced from hybrids resulting from crosses of S. acaule (2n=96) with different varieties of S. tuberosum (2n=48), 1967–1975

Combination	Yield per plant, g		Starch content, %	
	average	variations	average	variations
F_1 *S. acaule* × *S. tuberosum*				
ac × Pojana	150	10–1200	17.0	11–19.0
ac × MPI 50.140/5	120	50–1000	15.6	10–18.5
ac × Mariella	100	20–1000	16.0	10–18.0
ac × Khibinskii rannii	75	10–800	15.0	12–17.4
ac × Anoka	100	20–1000	16.5	11–18.4
ac × Carla	75	50–1000	15.0	10–17.6
B_1 (*S. acaule* × *S. tuberosum*) × *S. tuberosum*				
(ac × Priekul'skii rannii) × Earlaine	125	50–1500	17.5	11–19.8
(ac × Khibinskii rannii) × Kameraz	100	30–1200	16.8	10.9–18.0
(ac × Khibinskii rannii) × Earlaine	120	70–1200	16.3	10–18.6
(ac × Khibinskii rannii) Veselovskii	130	50–1300	15.4	11–17.8
(ac × Khibinskii rannii) × Mariella	120	50–1200	16.4	10–18.5
(ac × MPI 50.140/5) × MPI 50.140/5	200	70–1500	16.0	10–19.0
(ac × Gatchinskii) × Anoka	150	70–1200	15.8	11–17.5
(ac × Atzimba) × Atzimba	80	20–1000	15.0	10–17.0

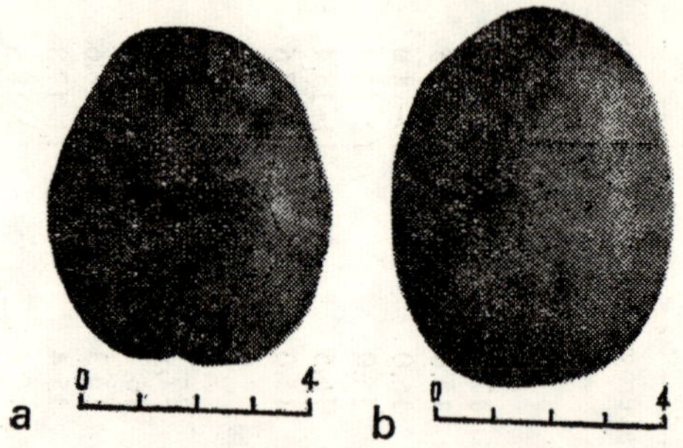

Fig. 5. Variation in the shape of tubers in the combination Priekul'skii rannii × (*S. punae* 96 × Priekul'skii rannii): a—tuber of irregular shape with deep eyes; b—tuber of good shape with shallow eyes.

involving the tetraploid *S. acaule*. The minimum yield was quite low—20–70 g, i.e. much lower than in backcrosses of tetraploid *S. acaule*. The average yield for different combinations was 80–200 g. The starch content fluctuates from 10 to 19.8%. The shape of the tuber was often bad, with deep eyes but occasionally it was better with shallow eyes (Fig. 5).

In concurrent backcrosses of hybrids with the involvement of the polyploid form of *S. acaule* the yield increases. Yield as well as starch content varies in hybrid combinations depending upon the components of *S. tuberosum*. The compact nature of the stolons improves considerably. The selection of forms valuable as regards their complex of characters is already possible among hybrids of this type.

The above-mentioned data show that in the case of concurrent hybridization of *S. acaule*, especially of its polyploid form, with diverse varieties of *S. tuberosum*, we can get practical results much more quickly than in preliminary hybridization of *Acaulia* with other wild species.

However, the latter method of involving *Acaulia* in distant hybridization should not be ignored, partly because it permits the combination of certain properties in hybrids which can only be taken from some wild species and their derivatives, particularly resistance to different dangerous diseases.

The preliminary crossing of wild species between themselves is much more effective when polyploid forms are included in the crossing. For example, according to our data, in the combinations (*S. acaule* 96 × *S. stoloniferum* 96) × *S. tuberosum* 48, (*S. acaule* 96 × *S. polytrichon* 96) × *S. tuberosum* 48 the yield varied from 200–1500; 100–1000; 50–800 g; respectively, i.e. approximately in the same way as in the combination (*S. acaule* × *S. tuberosum*) × *S. tuberosum*.

Resistance to some diseases and frost. Species of *Acaulia* are highly susceptible to Phytophthora. However, under field conditions hybrid plants from the crosses of species of this series with Phytophthora—resistant species *S. stoloniferum*, *S. polytrichon* and *S. vernei* are not infected by Phytophthora, except for some specimens with leaves which develop isolated necrotic spots without spore formation.

In subsequent crosses with *S. tuberosum*, a large number of plants in the combinations were infected by Phytophthora to various degrees. Nevertheless, even in complex hybrid combinations of Phytophthora-resistant species and *S. tuberosum* involving *Acaulia*, it was possible to single out hybrid plants comparatively resistant to Phytophthora under field conditions.

Phytophthora resistance in hybrids can be increased by including in crosses of *Acaulia*, Phytophthora-resistant species such as *S. demissum* or its derivatives. For example, hybrids of the cross (*S. acaule* × *Atzimba*) × *Atzimba* and others have field resistance to Phytophthora.

In this way, with the correct selection of components for hybridization with a species of series *Acaulia*, we can greatly increase the emergence of hybrids with field resistance to Phytophthora.

As already mentioned, we conducted large-scale hybridization between specimens of *Acaulia* resistant to virus X and specimens of *S. stoloniferum* and *S. polytrichon* resistant or super sensitive to viruses Y+A and resistant to Phytophthora. As a result we got hybrids with complex resistance to viruses X+Y+A (but susceptible to virus M). In the case of concurrent hybridization with *S. tuberosum* the number of plants with complex resistance to the three above-mentioned viruses was reduced drastically. However, even in complex trispecific hybrids, we could identify individual hybrids distinguished by their field resistance to the above-mentioned viruses (1, 4). Sometimes field resistance to leaf curling virus (L) is also inherent in hybrids of *S. acaule*.

Preliminary hybridization of *Acaulia* with some nematode-resistant specimens of a number of wild species, such as *S. bukasovii*, *S. multidissectum*, *S. spegazzinii*, *S. leptophyses*, *S. kurtzianum*, *S. famatinae*, *S. velascanum*, *S. vernei*, *S. microdontum*, *S. megistacrolobum*, *S. sanctae rosae*, *S. tarijense* and others (see Tables 1 and 5), has great significance in developing initial hybrid forms combining the valuable characters of *Acaulia* with resistance to different pathotypes of potato nematodes. Some of the above-mentioned species are characterized by frost resistance as well.

One of the important, but unsolved, problem is the development of agro-economically valuable and frost-resistant potato varieties. Highly frost-resistant specimens of species of *Acaulia* together with other species can be used as initial material for breeding with this aim.

At the first stage of hybridization with each species we observe high frost resistance in F_1. For example, in the F_1 of the combination *S. acaule* 48 × *S. tuberosum* 48 all hybrid plants withstood a frost of —4°C for a period of four hours in cold-rooms without any damage to their aerial parts (see Table 5). Regarding their habit, the plants in F_1 close to initial forms of *Acaulia* were distinguished by their rosette-like shape. Most of them do not develop tubers under long-day conditions. But if there are tubers they are very small; the yield does not exceed 100 g from one plant. Further crossings with *S. tuberosum* improve the yield and other economic properties of hybrid plants, but reduce their frost resistance. However, even among hybrids of concurrent backcrosses with *S. tuberosum*, we can isolate plants which can withstand frosts of −4°C for four hours with minor damage to their aerial parts.

Naturally high frost resistance is observed when frost-resistant varieties are crossed with each other, for example *S. acaule* × *S. toralapanum* and others. Frost resistance is reduced considerably when frost-susceptible wild species and varieties of *S. tuberosum* are used as components of crossing with *Acaulia*, as in the combination (*S. acaule* × *S. infundibuliforme)* × *S. tuberosum*.

To achieve the desired effect in breeding for frost resistance involving species of *Acaulia* and to combine frost resistance with resistance to Phytophthora, viruses and other valuable characters, it is necessary to greatly increase the scale of breeding work.

CONCLUSIONS

1. Wild South American tetraploid species of series *Acaulia* are almost never used in potato breeding, despite the fact that they are of considerable interest in breeding because there are specimens which are highly resistant to frosts and some viruses and which have other economically valuable characters.

2. Because of their phylogenetic remoteness, species of *Acaulia* are difficult to cross with species of other series. However, in the case of massive hybridization work it is possible to get viable seeds from crosses of *Acaulia* with many species of different ploidy levels relating to different series of South as well as North American groups, including the tetraploid *S. tuberosum*. In the majority of such cases, however, a very small quantity of seed develops per pollinated flower.

3. In crosses of tetraploid *Acaulia* with tetraploid components of crossing we often get tetraploid progeny. However, sometimes, because of the presence of nonreducing gametes of *Acaulia*, the hybrids developed are not tetraploid but hexaploid (2n = 72). The progeny from crosses of *Acaulia* (2n = 48) with diploid species are mostly triploid. Tetraploids are found rarely.

4. Use of artificially developed polyploid forms of *Acaulia* (2n = 96) for interspecific crosses improves crossability and increases the quantity of seeds per pollinated flower. Crossing with octoploid *Acaulia* gives segregation of the chromosome number from 48 to 72.

5. As a rule, recurrent hybridization of *Acaulia* with *S. tuberosum* improves crossability, particularly when octoploid *Acaulia* is used to get the primary hybrid.

6. At present it is not impossible to involve species of *Acaulia* in breeding

work. The breeding work can be carried on by developing bispecific hybrids from concurrent crosses of *Acaulia* with *S. tuberosum* or through preliminary hybridization between *Acaulia* and other valuable species and recurrent crossings of these hybrids with *S. tuberosum* or through crossings of the hybrids of *Acaulia* between themselves (concurrent backcrosses with *S. tuberosum*) and hybrids also of concurrent backcrosses with *S. tuberosum* and other species.

7. Bispecific hybridization involving *Acaulia* and *S. tuberosum* is more promising in obtaining high yielding hybrids with a complex of agro-economically valuable characters within comparatively short periods of time. Trispecific hybridization is more complicated, and requires considerably more time, but in the end it can be more effective, particularly for developing forms with resistance to various diseases.

REFERENCES

1. Bukasov, S.M. and A.Ya. Kameraz. 1972. Selektsiya i Semenovodstvo Kartofelya (Breeding and Seed Production of the Potato). Leningrad, 359 pages.
2. Vavilova, M.A. and N.A. Zhitlova. 1973. Uroven' ploidnosti gibridov *Solanum acaule* Bitt. (The ploidy level of hybrids of *Solanum acaule* Bitt.). *Genetika*, Vol. IX, No. 11, pp. 32–37.
3. Dremlyug, L.A. 1937. Troinoi morozostoikii gibrid *S. acaule* Bitt. × *S. tuberosum* × *S. tuberosum* (A triple frost-resistant hybrid of *S. acaule* Bitt. × *S. tuberosum* × *S. tuberosum*). *Dokl. AN SSSR*, Vol. 16, No. 8, pp. 431–434.
4. Kameraz, A.Ya. and V.N. Ivanova. 1971. Selektsiya kartofelya na ustoichivost' k mozaichnym virusam (Potato breeding for resistance to mosaic viruses). *Trudy Po Prikl. Bot., Gen. i Sel.*, Vol. 44, No. 1, pp. 116–143.
5. Correll, D.S. 1962. The potato and its wild relatives. Renner, Texas, 606 pages.
6. Hermsen, J.G. 1966. Crossability, fertility and cytogenetic studies in *S. acaule* × *S. bulbocastanum*. *Euphytica*, Vol. 15, No. 2, pp. 149–153.
7. Irikura, Y. 1968. Studies on interspecific crosses of tuber-bearing Solanums. I. Overcoming crossing compatibility between *S. tuberosum* and other *S.* species by means of induced polyploids and haploids. Reprint from the Research Bull., No. 92, pp. 21–38.
8. Lamm, R. 1943. Notes on an octoploid *Solanum punae* plant. *Hereditas*, 29, pp. 193–194.
9. Propach, H. 1938. Kreuzbarkeit von Solanum-Arten untereinander und mit Kulturkartoffeln und die Fertilitat der Bastrade. Forschungsdienst, 6, pp. 273–282.

10. Swaminathan, M.S. 1950. Einige Verfahren für die Verwendung wilder Solanum Arten zu Zuchtzwecken. Züchter, Vol. 20, pp. 28–34.
11. Swaminathan, M.S. 1951. Notes on induced polyploids in the tuber-bearing *Solanum* species and their crossability with *S. tuberosum*. *Am. Potato J.*, Vol. 28, pp. 16–24.
12. Ugent, D. 1966. Hybrid weed complexes in *Solanum*, section *Tuberarium*. Wisconsin, 123 pages.

TRUDY PO PRIKLADNOI BOTANIKE, GENETIKE I SELEKTSII

1978 Vol. 62, No. 1

Use of the Wild Diploid Species of the Potato *S. chacoense* Bitt. in Interspecific Hybridization

A.Ya. Kameraz
Doctor of Agric. Sciences

N.A. Zhitlova, V.N. Ivanova
Candidates of Agric. Sciences

The polymorphous diploid (2n=24) potato species S. chacoense Bitt., which is little used in breeding work, comprises forms with numerous valuable characters, including resistance to viruses, scabs—Streptomyces scabies (Thaxter) Waksman et Henrici and Spongospora subterranea (Wallr.) Lagerh, various pathotypes of cancer, black leg, nematodes—Heterodera rostochiensis Wollenweber and Ditylenchus destructor Thorne, Colorado potato beetle, Epilachna viginti octo maculata. These samples are also characterized by a higher content of dry matter in the tubers.

Usually S. chacoense crosses poorly with other species. At the same time it was successfully involved in hybridization with various species at different ploidy levels, belonging to various series. Artificial polyploidization of S. chacoense greatly improves its crossability with S. tuberosum.

Among the hybrids of S. chacoense obtained from second crossings with S. tuberosum, certain forms may be singled out for their productivity and increased starch content and for the taste qualities of the tubers.

THE IMPORTANCE OF S. CHACOENSE AS INITIAL MATERIAL FOR BREEDING

Until recently a comparatively small number of wild species was used for breeding new varieties of the potato throughout the world. At the same time, the exceptional diversity of wild species was well known and some of them, according to available data, have quite valuable properties and are of considerable interest as initial material for breeding. There is no doubt that such species include the polymorphic diploid (2n = 24) of *S. chacoense* Bitt. According to the system developed by S.M. Bukasov (1974), it belongs to subsection *Orientale* Buk. of series *Glabrescentia* Buk. (Fig. 1).

This species of the potato is found extensively in the northern part of Argentina, in the adjoining parts of Paraguay and in southwest Brazil, at an altitude ranging from 500 to 2500 m above mean sea level, beginning at the eastern slopes and foothills of the Andes up to the arid plain of Chaco. S.M. Bukasov [1] mentions that the relief of this geographic range is not as dissected as in the hill zone of the Andes. Therefore, there are no clear differences between species which were earlier described by G. Bitter, E. Chasler and S.V. Yuzepchuk and S.M. Bukasov: *S. chacoense* Bitt., *S. subtilius* Bitt., *S. schickii* Juz. et Buk., *S. knappei* Juz. et Buk., *S. emmeae* Juz. et Buk., *S. gibberulosum* Juz. et Buk. and *S. boergeri* Buk. It must be mentioned that all these species are easily cross-pollinated among themselves, not only under natural conditions in their native habitat but also under cultivation in other geographic regions including in the outskirts of Leningrad and it makes their exact classification a difficult task.

Several times we have emphasized the need to use *S. chacoense* in practical breeding work [1, 3–7 and others]. On the whole, this species is distinguished by its field resistance to different viruses, since under artificial infection conditions we can isolate specimens resistant to viruses X, Y and A and leaf curling virus. Individual specimens have resistance to common and powdery scab, different pathotypes of potato wart, black-leg, potato and stem nematodes, Colorado potato beetle and 28-spotted potato bug (Epilachne) and they also have a higher content of dry matter in their tubers. This species is very promising in breeding for increasing the starch content of tubers.

As shown by our studies conducted at the Pushkin Laboratories of the VIR (All-Union Institute of Plant Industry), within the limits of this species there is great diversity of these and other economically important characters. Therefore, before its inclusion in a breeding program, we must isolate specimens which have the desired characters or complex of characters. In a number of cases it is appropriate to conduct a preliminary cross between different forms of the species in order to develop initial specimens for further hybridization which had the maximum number of economically valuable characters that are desired to be transmitted to the hybrid progeny.

Fig. 1. Wild species of the potato *S. chacoense* Bitt. (2n=24). Plants:
a—specimen resistant to virus Y; b—hybrid from recurrent crossings with *S. tuberosum.*

The serious shortcomings of all specimens of *S. chacoense* are high susceptibility to Phytophthora, long stolons and bitter taste of the tubers. These negative characters are inherited by hybrid progeny even in backcrosses of different degrees of complexity. At the same time, with the proper selection of plants of *S. tuberosum*, these shortcomings can be removed in recurrent backcrosses.

THE HYBRIDIZATION OF S. CHACOENSE WITH OTHER SPECIES OF THE POTATO

As a rule, diploid *S. chacoense* is difficult to cross with other species and even when hybrid berries do form it gives a very small quantity of seeds. At the same time, if it is used as the female parent, positive results are obtained in crosses with different diploid species belonging to different series: *Berthaultiana* Buk., *Transaequatorialia* Buk., *Andigena* Buk., *Simpliciora* Buk., *Alticola* Buk., *Megistacroloba* Card. et Hawk., *Conicibaccata* Buk., *Tuberosa* (Rydb.) Buk. (*S. tuberosum—dihaploid*), *Tarijensa* Corr., *Yungasensa* Corr., *Commersoniana* Buk., *Demissa* Buk., *S. verrucosum* Schlechtd., and also with species of other ploidy levels: *S. demissum* Lindl.—hexaploid (series *Demissa*), *Longipedicellata* Buk. (*S. stoloniferum* Schlechtd., *S. polytrichon*—both tetraploids). The polyploid form of *S. chacoense* was also successfully crossed with diploid *S. infundibuliforme* Phil. (series *Cuneoalata* Hawk.).

When the diploid *S. chacoense* was used as the male parent in crosses with some species of almost all the above-mentioned series, we got hybrid berries with a small quantity of seeds. Berries also developed as a result of pollination of tetraploid *S. acaule* Bitt. (series *Acaulia* Juz.) and triploid *S. vallis mexici* Juz. (series *Longipedicellata* Buk.). From what has been stated above, it shows that a positive result was obtained mostly from crosses of diploid *S. chacoense* with other diploid species. However, berries sometimes developed in hybridization with species of other ploidy levels.

The crossing of *S. chacoense* with different wild species has no direct practical importance because it does not lead to any improvement in the economically valuable characters of the hybrids. It can be considered only as an intermediate stage on the way to crossings with the improved varieties without which it is impossible to get agro-economically valuable forms through interspecific hybridization. Direct crossing of wild species with *S. tuberosum* has a more practical significance.

Diploids of *S. chacoense*, when crossed with *S. tuberosum*, produce mostly tetraploids, very rarely triploids in the F_1 (10–14). The tetraploid generation has much more vigor than the triploids and is distinguished by the high viability of the pollen [3].

Crosses of different specimens of diploid *S. chacoense* with different varieties of *S. tuberosum* (Fig. 2) showed that use of diploid *S. chacoense* as

Table 1. The results of crosses of diploid specimens of S. chacoense (2n=24) with varieties of S. tuberosum (2n=48), 1964–1973

Combination	Number of pollinated flowers	Successful crosses, %	Number of berries		Number of seeds		
			Total	With seeds	Total	Per berry	Per pollinated flower
F_1 ch × t	1104	4.4	328	49	92	1.8	0.083
F_1 t × ch	171	17	69	30	53	1.7	0.3
F_1 ch × t (2n=24)	29	—	—	—	—	—	—
F_1 t (2n=24) × ch	30	53	19	16	50	3	1.6
B_1 (ch × t) × t	1030	35	416	378	40996	108	40.0
B_1 t × (ch × t)	394	66	268	261	36186	139	92
B_1 [(ch × t) × t] × t	1286	21	279	266	29711	112	23
B_1 [t × (ch × t)] × t	1256	16	221	206	31203	151	25
B_1 t × [t × (ch × t)]	639	46	301	298	32651	109	51
B_1 t × [(ch × t) × t]	162	42	71	69	5287	76	33
B_1 {[t × (ch × t)] × t} × t	248	25	61	61	8400	138	34
B_1 {[(ch × t) × t] × t} × t	336	44	175	150	22810	152	67

Note: Here and later ch=*S. chacoense*, t=*S. tuberosum*.

the female as well as the male parent gives a very small percentage of berries with seeds (Table 1). The average quantity of seeds per berry was 1.7–1.8 and per pollinated flower 0.083 (*S. chacoense* as the female parent) and 0.3 (*S. chacoense* as the male parent). In the latter case the percentage of successful crossings also increases. However, it is necessary to take into consideration the fact that there were far fewer pollinated flowers in this variant. In the context of the small quantity of pollinated flowers, the percentage of successful crosses increased when plants of diploid *S. tuberosum* were taken as the female parent; in this case there was also some increase in the average quantity of seeds per berry [3] and per pollinated flower [1, 6]. When dihaploid *S. tuberosum* was used as the male parent we did not get hybrid berries.

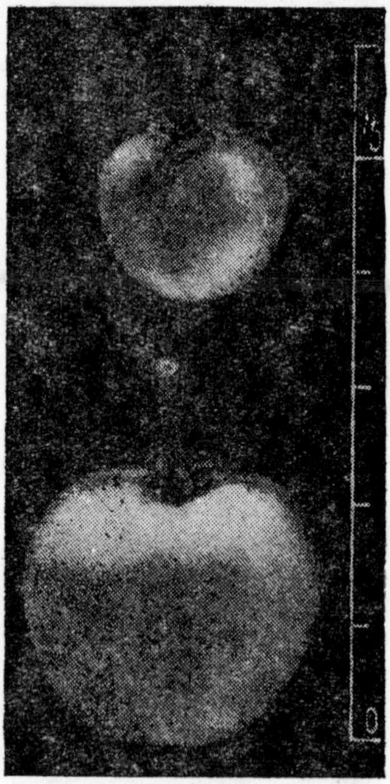

Fig. 2. Berries of *S. chacoense*, diploid from (above) and polyploid form (below).

In different versions of subsequent crosses of the F_1 hybrids of *S. chacoense* with *S. tuberosum*, there is a considerable increase in the percentage of successful crosses and the average quantity of seeds per berry and per pollinated flower. In backcrosses of the type (*S. chacoense* × *S. tuberosum*) × *S. tuberosum* and *S. tuberosum* × (*S. chacoense* × *S. tuberosum*) the quantity of seeds per berry was 108–139 and per pollinated flower 40–92. The average quantity of seeds in the berry increases even in recurrent backcrosses of different degrees of complexity.

The phylogenetic remoteness of the species being crossed significantly affects the percentage of hybrid berries with seeds and the quantity of seeds per berry and per pollinated flower in the primary crosses. The results of the crosses in this case depend comparatively little on the peculiarities of the varieties of *S. tuberosum*. However, in backcrosses and in recurrent backcrosses of different degrees of complexity, the peculiarities of the varieties assume great importance and this affects the results of hybridization to a great extent.

In the primary crosses, particularly when *S. chacoense* is taken as the female parent, a large number of the berries (85%) were found to be parthenocarpic. In backcrosses and recurrent backcrosses the quantity of such berries is reduced drastically, but nevertheless such berries are present in all versions of the crossings.

Preliminary polyploidization of *S. chacoense* increases the size of berries (Fig. 2) and sharply improves its crossability with *S. tuberosum*. This becomes evident when wild species are used as the female parent and, to an even greater extent, when they are used as the male parent (Table 2). Here the quantity of hybrid berries with seeds is as high as 70.7–96.5%, the quantity of seeds per berry 125–136 and per pollinated flower 98.2–131.7. An interesting fact here is that in the hybridization of diploid *S. chacoense* with *S. tuberosum* large quantities of parthenocarpic berries are formed, while they are absolutely absent if we use the tetraploid *S. chacoense*. All hybrid berries thus formed are found to have fully viable seeds.

In the work of K.A. Kotova [8] crosses involving the tetraploid form of *S. chacoense* as the male parent were found to be the most promising. In this case 40.2% of the crosses were successful. However, in contrast to our data (Table 2), the quantity of seeds per pollinated flower was very small—20.8 (in our experiment it was 131.7). It is possible that Kotova had a low number of primary crosses and as a component of crossing she used varieties such as Oda, Greta, Saphir, Rector, Adwira and Veselovskii 2–4, which were less suitable for hybridization with tetraploid *S. chacoense*. According to Kotova's data, the pollen of *S. chacoense* ($2n=48$) was highly fertile: 51.9% of the pollen grains were colored with acetocarmine and 22.8% germinated on the nutritional medium of T. Plekhanova [9].

Subsequent crossings with *S. tuberosum*, in comparison with the primary crossing of polyploid forms, do not increase either the percentage of berries formed with seeds, or the average quantity of seeds per berry and per pollinated flower (Table 2). These indices obviously change depending upon the peculiarities of the varieties of *S. tuberosum* involved in the crosses.

In our experiments on crosses of the type *S. tuberosum* × (*S. chacoense* × *S. tuberosum*) involving the polyploid form, 30.5% of the crosses were successful; the quantity of seeds per berry was 117 and per pollinated flower 36 (Table 2). In similar crosses, but involving the diploid form of *S. chacoense*, the

Table 2. The results of primary and concurrent crosses of the polyploid form of S. chacoense (2n=48) with S. tuberosum (2n=48), 1964–1973

Combination	Number of pollinated flowers	Successful crosses, %	Number of berries		Number of seeds		
			Total	With seeds	Total	Per berry	Per pollinated flower
ch × t	423	70.7	327	327	40875	125	98.2
t × ch	259	96.5	251	251	34136	136	131.7
t × (ch × t)	190	30.5	78	58	6817	117	36
(ch × t) × t	191	29.8	82	57	4386	77	23
t × (t × ch)	94	57.4	60	54	6046	112	64
(t × ch) × t	125	28.0	46	35	2365	67	19

Table 3. Results of the crossings of F_1 hybrids of S. chacoense (2n=48) with different varieties of S. tuberosum (1973–1974)

Combination	Number of pollinated flowers	Successful crosses, %	Number of berries		Number of seeds		
			Total	With seeds	Total	Per berry	Per pollinated flower
MPI 50.140/5 × (ch × Amsel)	57	56	32	32	3048	95	53.5
Bake-King × (ch × Amsel)	42	21	9	9	890	99	21.7
Chieftain × (ch × Amsel)	25	16	4	4	483	120	19.3
(ch × Kameraz) × Amsel	45	31	14	14	550	40	12.2
(ch × Kameraz) × MPI 50.140/5	25	48	12	9	595	66	23.8
(ch × Carla) × Amsel	25	52	13	11	712	65	28.5
(ch × Carla) × MPI 50.140/5	45	58	26	16	1434	90	32.1
(ch × MPI 50.140/5) × Amsel	25	48	12	9	1668	185	67
(ch × Atzimba) × MPI 50.140/5	45	53	24	19	1706	90	38
(ch × Amsel) × Fidelio	20	55	11	7	332	47	16.6
(ch × Amsel) × Agro	31	52	16	9	186	20	6
(ch × Amsel) × Arensa	12	33	4	3	102	34	8.5
(ch × Amsel) × Epoka	23	52	12	7	280	40	12.3

corresponding indices were as follows: 66%, 139 and 92 (see Table 1). In the crosses (*S. chacoense* × *S. tuberosum*) × *S. tuberosum* involving both the polyploid and diploid forms of *S. chacoense* the indices were accordingly: 29.8 and 35%; 77 and 108 seeds; 23 and 40 seeds. However, a final conclusion should not be drawn from these data because in order to do so, we should have had in each treatment an approximately equal quantity of pollinated flowers and should have conducted crossings with the same specimen of the wild species and with similar varieties of *S. tuberosum*.

Notwithstanding this, we can draw the following conclusions from the above data, and these should be taken into consideration when involving *S. chacoense* in interspecific hybridization.

1. Induced polyploidization of *S. chacoense* sharply improves its crossability with *S. tuberosum*, as expressed by the increase in the percentage of hybrid berries with seeds as well as the quantity of seeds per berry and per pollinated flower.

2. Subsequent hybridization of F_1 with *S. tuberosum* can be successfully conducted by involving the diploid as well as the tetraploid forms of *S. chacoense* in the primary crosses. In these cases, as well as in backcrosses and recurrent backcrosses, a large number of berries with fully viable seeds are developed and the average quantity of seeds per berry and per pollinated flower is also high. This ensures a firm basis for a further breeding program with hybrid material.

3. The success of hybridization involving diploid as well as tetraploid forms of *S. chacoense* in the primary crosses is linked with the peculiarities of the varieties of *S. tuberosum* being used as breeding plants and with the hybrids of F_1 (Table 3). In some backcrosses careful choice of the parents ensured a high efficacy in the crossings. For example, in the combination (*S. chacoense* 2n=48 × MPI 50.140/5) × Amsel the quantity of hybrid seeds per berry was 185 and per pollinated flower 67; in the combination MPI 50.140/5 × (*S. chacoense* × Amsel) it was 95 and 53.5, while in the combination (*S. chacoense* × Amsel) × Agro, it was only 20 and 6, and so on.

The fertility of the pollen of F_1 hybrids from crosses of *S. chacoense* (2n = 48) with varieties of *S. tuberosum* was determined by means of coloring pollen grains with acetocarmine as well as by discovering if they germinate on a sugar-agar agar medium (Table 4). From the data presented here it is evident that hybrids of this type have a good quality of pollen and can be successfully used in further hybridization. This is testified to by the results of crossings as well (Tables 2 and 3).

K.A. Kotova [8], on the basis of a comparatively small number of crosses, also comes to a similar conclusion. According to her data, the use of F_1 hybrids of *S. chacoense* as the female parent in further hybridization with *S. tuberosum* gives a much higher percentage of successful crosses (20.8 against 9.4). However, the berries obtained through involvement of hybrids of *S.*

Table 4. The fertility of pollen in hybrids in the F_1 from crosses of S. chacoense (2n=48) with different varieties of S. tuberosum (1973–1974)

Combination	Colored pollen grains, %	Germinated pollen grains, %
ch × Amsel	78.7	58.6
ch × Kameraz	64.5	69.0
ch × Carla	81.0	78.0
ch × MPI 50.140/5	90.0	84.6
ch × Atzimba	84.2	72.6

chacoense as the male parent had a larger number of seeds (the average number of seeds in one berry was 124 against 64); the quantity of hybrid seeds per pollinated flower was almost the same (13.27 and 11.7), but they were much smaller than in our experiment (Tables 2 and 3).

YIELD AND STARCH CONTENT OF TUBERS FROM HYBRIDS OF S. CHACOENSE (2n=24)

Sometimes in hybrid saplings of different degrees of complexity, from crosses with diploid specimens of *S. chacoense*, some plants are distinguished by a comparatively good yield of tubers (Table 5). In the F_1 the yield varied from 5 to 250 g. The starch content of tubers of individual saplings reached 23%.

In the F_2 of *S. chacoense* (2n = 24) × *S. tuberosum* the average maximum yield increased sharply, reaching 1200 g per plant. Even more significant was the maximum yield in backcrosses such as (*S. chacoense* × *S. tuberosum*) × *S. tuberosum*, which was 1690 g; the starch content of some plants was as high as 25%.

In further complicated recurrent crossings with different varieties of *S. tuberosum* (two-three backcrosses) the average and the maximum yields of plants do not always increase. In such cases the yield depends not so much on the number of recurrent crossings, as on the peculiarities of the plants of the cross, i.e. on the varieties of *S. tuberosum*. Similarly, whether the plants are used as the male or the female parent in the combination is also known to have a certain importance. Barring a few exceptions, in recurrent backcrosses we observe a tendency for the average yield to be higher than with backcrosses. This takes place because of a reduction in the size of variations in the given character in the families of recurrent backcrosses: often a minimum yield increases considerably in the case of recurrent backcrosses.

Among diverse hybrids of *S. chacoense*, our attention is drawn to the presence of individual saplings with a high starch content in the tubers (Table 5).

In all cases of similar crosses, but using the polyploid form of *S. chacoense*,

Table 5. Some economic properties of tubers of hybrid saplings from the crosses of S. chacoense specimens with varieties of S. tuberosum (2n=48), 1968–1974

Combination	Number of saplings	Yield from one plant, g		Starch content, %	
		average	variation	average	variation
F_1 ch × t	46	65	5–250	19	12.5–23
F_2 ch × t	9	611	30–1200	16.2	12.7–18.8
F_1 ch 48 × t	9	311	90–960	16.2	10.5–18.5
(ch × t) × t	174	405	20–1690	16.7	<10–25
(ch 48 × t) × t	50	480	80–1900	16.6	<10–23.8
t × (ch × t)	108	409	50–1200	16.9	<10–23.8
t × (ch 48 × t)	32	745	290–1320	12.7	10–20
(t × ch) × t	11	44	2–220	16.1	12.9–18.8
(ch × t) × t × t	128	540	80–1460	17.5	10.5–25
t × (ch × t) × t	4	557	220–970	12.5	<10–14.4
(t × ch) × t × t	12	518	140–700	14.5	13.6–16.3
[t × (ch × t)] × t	60	395	90–1000	17.8	11.5–23.3
t × [t × (ch × t)]	34	680	280–1440	15.2	<10–21
t × {[(t × ch) × t] × t} × t	5	574	350–880	14.0	12.9–15.7
{[(ch × t) × t] × t} × t	12	325	130–510	18.7	15.4–25
{[t × (ch × t)] × t} × t	40	422	100–780	17.6	<10–22.5

Note: ch=*S. chacoense* (2n=24); ch 48=*S. chacoense* (2n=48).

the average as well as the maximum yield of saplings was much higher; this was not observed in respect of the starch content.

After eliminating the undesirable ones from the previous year, the quantity of clones is reduced considerably in tuber production in all combinations. Naturally, the average yield increased considerably because during the selection of saplings the less yielding ones were rejected. The maximum yield in some cases was increased in comparison with the saplings, while in others it was decreased. The maximum starch content was often high—up to 20–26%. In tuber production, as well as in saplings, we observed a vast range of variability regarding the starch content in the tubers.

On the basis of an analysis of the saplings and tuber production of hybrids from diverse crosses of the diploid form of *S. chacoense* with different varieties of *S. tuberosum*, it follows that among recurrent backcrosses we can isolate a considerable number of high yielding clones with increased starch content of the tubers. The problem for the plant breeder lies in the fact that, generally, long stolons, which are responsible for creating an undesirable spreading in the cluster of tubers, are inherent in hybrids of *S. chacoense* even after a number of recurrent crossings with *S. tuberosum*. Subsequent hybridization, first of all, must be aimed at removing this negative character, while maintaining in the hybrids the complex of desirable properties including good yield and increased starch content in the tubers.

YIELD AND STARCH CONTENT OF TUBERS FROM HYBRIDS OF S. CHACOENSE (2n=48)

As was already mentioned, in comparison with the diploid, the crossability of the polyploid *S. chacoense* with different varieties of *S. tuberosum* increases sharply. A greater number of hybrid seeds becomes available to the plant breeder permitting him to increase the scale of breeding work. Besides, the maximum and average yield of the hybrids increases significantly in the first generations and in backcrosses (Fig. 3).

Fig. 3. A cluster of tubers from the hybrid plant of (*S. chacoense* 2n=48 × Carla) × Khibinskii rannii.

In order to elucidate this position, we present in Table 6 data for 1972 on the yield and starch content in the first tuber production of some hybrids obtained from crosses with the polyploid form of *S. chacoense*. From each combination we planted hybrids selected from 1971 saplings free from virus infection and distinguished by shorter stolons and producing a good yield of tubers in the field under natural day length conditions. These results show that almost all plants of *S. tuberosum* used by us in crossing with the tetraploid *S. chacoense* in F_1 gave similar results regarding the yield of the tubers. Better indices were obtained when a hybrid of MPI 50.140/5 was used as the female as well as the male parent. In this case the maximum yield of individual hybrids from the combination was as high as 1200–1500 g, with an average yield of 180–200 g. With the involvement of the variety Chieftain the maximum yield increased to 1200 g and the average yield to 220 g.

Table 6. Some economic properties of the first tuber production of hybrids from crosses of S. chacoense (2n=48) with different varieties of S. tuberosum (1972)

Combination	Yield from one plant, g		Starch content of tubers, %		Average number of tubers in the cluster
	average	variation	average	variation	
ch × Kameraz	160	0–1000	18.3	14.2–21.3	4.2
ch × Carla	180	20–1000	19.2	15.6–23.0	6.1
ch × Atzimba	60	0–700	17.6	13.0–21.3	12.0
ch × Anoka	90	30–900	20.6	14.5–22.1	9.0
ch × MPI 50.140/5	180	46–1200	18.4	13.0–21.0	6.0
ch × Mars	90	28–1000	19.0	14.0–20.6	7.3
ch × Chieftain	180	40–1000	21.9	16.3–24.3	6.1
ch × Bake-King	160	50–950	20.4	16.0–21.8	7.0
ch × Argo	150	32–1000	18.6	14.0–20.4	8.3
ch × Amsel	100	0–900	17.8	14.3–19.2	4.0
Chieftain × ch	220	46–1200	19.8	16.3–23.0	6.2
Bake-King × ch	190	38–1000	19.0	16.0–21.0	7.3
MPI 50.140/5 × ch	200	40–1500	17.0	14.3–19.3	6.1
Atzimba × ch	100	0–900	17.1	13.8–18.9	8.6
Carla × ch	120	60–1000	18.6	15.0–21.8	6.4
(ch × Kameraz) × Amsel	300	180–2000	18.0	11.6–20.0	6.2
(ch × Carla) × Amsel	350	150–2000	18.1	12.7–19.8	8.6
(ch × MPI 50.140/5) × Carla	420	200–2500	17.3	13.0–19.0	8.0
(ch × Amsel) × Argo	350	100–2000	16.4	11.0–18.6	8.3
(ch × Amsel) × Bake-King	650	300–2600	17.6	14.2–18.0	7.3
(ch × Carla) × MPI 5.140/5	550	300–2500	17.8	13.1–19.0	8.0
(ch × Atzimba) × Carla	300	150–2000	16.7	10.7–18.1	7.8
(ch × Atzimba) × Anoka	350	120–1800	17.0	11.4–19.0	8.0
(ch × Atzimba) × Earlain	320	140–1500	15.9	12.1–17.0	9.0
(ch × Mars) × Fidelio	410	160–2000	16.7	11.0–18.6	6.7
(ch × Amsel) × Mars	370	180–2000	18.0	12.0–19.0	8.1
(ch × Kameraz) × Prudor	320	130–1800	18.4	13.3–20.0	6.4

A higher content of starch in individual clones is also noteworthy. It was as follows in combinations involving these varieties: Chieftain—up to 24.3% (average for the combination—21.9%), Carla—23.0% (19.2%), Anoka—22.1% (20.6%) and so on.

The yield increases dramatically in backcrosses and was as high as 2500–2600 g in individual clones. In such cases the average yield of these combinations was 550–650 g per plant.

In our combinations of backcrosses the starch content was lower than in the F_1. The maximum content of starch in individual clones reached 20% while the average starch content was 18.4%.

SOLANINE CONTENT IN TUBERS OF HYBRIDS OF S. CHACOENSE

According to some data, specimens of *S. chacoense* as well as its hybrids are distinguished by an increased content of solanine in the tubers. This is confirmed by the results of a biochemical analysis within the combination of the hybrids made by us (Table 7). At the same time, sometimes even in the F_1 and F_2 as well as in backcrosses, the solanine content was not higher than in some varieties obtained from intraspecific hybridization in *S. tuberosum*. Anyway, a random analysis of hybrids of *S. chacoense* (2n = 24 and 48) testifies to the possibility of selecting forms from those in which the solanine content in the tuber does not exceed the average range.

Table 7. Solanine and dry matter contents in the tubers of hybrids of S. chacoense and varieties of S. tuberosum (1973)

Hybrids and varieties	Dry matter, %	Solanine, mg/100 g
Interspecific hybrids		
F_1 ch × Kameraz	23.83	75.5
F_2 ch × Kameraz	24.30	86.0
F_1 ch × Carla	23.93	56.6
F_2 ch × Carla	25.00	7.72
F_1 ch × Atzimba	22.85	7.72
F_2 ch × Atzimba	17.30	30.0
Carla × F_1 (*S. chacoense* 48 × Amsel)	24.10	29.2
Bake-King × F_1 (*S. chacoense* 48 × Amsel)	25.13	Traces
F_2 ch 48 × Atzimba	26.85	123.6
F_1 ch × MPI 49.540/2	24.05	109.9
B_1 ch 24 × 3070 (d) × Gunda	19.72	22.3
B_1 [(Spekula × ch 24) × (Allerfrüheste Gelbe)] × Arensa × Gunda	19.60	12.6
B_1 [(Atzimba × (ch 24 × Arensa)] × Grata	20.25	41.2
Varieties of S. tuberosum		
Druzhba	17.10	20.6
Arran Banner	20.25	Traces
Claustar	16.20	14.9
Jette	22.50	38.9
Edda	22.70	0.0

We conducted an extensive tasting program of tubers of different complex hybrids involving *S. chacoense*. This enabled us to select a large quantity of tubers from them that had a good taste and a pith that did not darken.

RESISTANCE OF HYBRIDS OF S. CHACOENSE TO PHYTOPHTHORA AND OTHER DISEASES

As already mentioned above, one of the serious shortcomings of *S. chacoense* is the high susceptibility of all its specimens to Phytophthora. Therefore, at all stages or in the concluding stage of hybridization with this species it is necessary to use Phytophthora-resistant plants. It follows from our numerous data that it is possible to get a large quantity of hybrids with high resistance to Phytophthora, even in years of severe infection.

In this context it is worth noting the data of 1974, when a strong and early outbreak of Phytophthora was observed under the conditions in Leningrad province. Many races were observed in the field population beginning from the simple [1; 2; 3; 4] to the complex, including the race 1.2.3.4. Under these conditions plants of almost all the improved varieties were completely infected by Phytophthora by the time of harvesting; however, some complex hybrids obtained from a Phytophthora-resistant wild species and *S. chacoense* did not have any symptoms of the disease. Hybrids highly resistant to Phytophthora were obtained in combinations involving the following species: *S. demissum, S. chacoense, S. tuberosum*; *S. demissum, S. stoloniferum, S. chacoense, S. tuberosum*; *S. demissum*; *S. vallis mexici, S. chacoense, S. tuberosum*; *S. demissum, S. vernie, S. chacoense, S. tuberosum*; *S. stoloniferum, S. chacoense, S. tuberosum* and others. As a rule, in complex combinations varieties of *S. tuberosum* were involved 2–3 times or more in the crosses. Sometimes forms of *S. andigenum* were also used together with *S. tuberosum*. In some cases all hybrids of the combinations, and in others 50–80% of them, were free from symptoms of Phytophthorasis infection.

Among hybrids of *S. chacoense* we noticed common scab and virus infection to a small extent. What has been stated above undoubtedly testifies to the fact that *S. chacoense* should be utilized more extensively in potato breeding than it has been up till now.

CONCLUSIONS

1. The polymorphic diploid species of the potato *S. chacoense* Bitt., practically unused in potato breeding, produces specimens distinguished by different valuable characters including resistance to common and powdery scab, different pathotypes of potato wart, black-leg, potato and stem nematodes, Colorado potato beetle, 28-spotted potato bug and also by an increased content of

dry matter in the tubers. One of the shortcomings of this species is its high susceptibility to Phytophthora.

2. Hybrids were obtained from crosses of *S. chacoense* with different potato species. Induced polyploidization of *S. chacoense* sharply improves its crossability with all plants, including those crossed with the tetraploid *S. tuberosum*.

3. Among hybrids of *S. chacoense* from recurrent crosses with *S. tuberosum* we can isolate forms distinguished by high yield, increased starch content and high taste qualities of the tubers.

4. Besides *S. chacoense* and varieties of *S. tuberosum*, inclusion of Phytophthora-resistant plants in the complex hybrid combination provides an opportunity to get a large number of Phytophthora-resistant hybrids in the combination; these are simultaneously characterized by other agro-economically valuable characters.

5. Results of the study of initial forms and hybrids of *S. chacoense* suggest the need for its extensive use in practical breeding programs.

REFERENCES

1. Bukasov, S.M. and A.Ya. Kameraz. 1972. Selektsiya i Semenovodstvo Kartofelya (Breeding and seed production of the potato). Izd-vo "Kolos", Leningrad, 359 pages.
2. Bukasov, S.M. 1974. Morfologiya Kartofelya (Morphology of the potato). *Trudy Po Prikl. Bot., Gen. i Sel.*, Vol. 53, pp. 3–33.
3. Kameraz, A.Ya. 1959. Perspektivy ispol'zovaniya dikikh vidov kartofelya v selektsii (Prospects of using wild potato species in breeding). *Kartofel'*, No. 1, pp. 38–43.
4. Kameraz, A.Ya. 1972. Selektsiya kartofelya na ustoichivost' k naibolee vredonsonym i rasprostranennym boleznyam i vreditelyam (Potato breeding for resistance to the most virulent and common diseases and pests). In the book: *Selektsiya i Semenovodstvo Kartofelya, Ovoshchnykh, Plodovykh Kul'tur i Vinograda*. Moscow, pp. 29–38.
5. Kameraz, A.Ya. 1973. Mezhvidovaya i vnutrividovaya gibridizatsiya kartofelya (Interspecific and intraspecific hybridization of the potato). In the book: *Genetika Kartofelya*, Moscow, pp. 104–121.
6. Kameraz, A.Ya. 1973. Mezhvidovaya gibridizatsiya v selektsii kartofelya (Interspecific hybridization in potato breeding). *Sel'skokhozyaistvennaya Biologiya*, Vol. VIII, No. 5, pp. 643–651.
7. Kameraz, A.Ya. 1974. Itogi i perspektivy selektsii kartofelya na ustoichivost' k virusam (Results and prospects of potato breeding for resistance to virus). In the book: *Virusy i Virusnye Boleznii Rastenii*. Kiev, pp. 111–115.
8. Kotova, K.A. 1974. Ispol'zovanie v selektsii Kartofelya dikogo vida S.

chacoense Bitt. (Use of the wild species *S. chacoense* Bitt. in potato breeding). In the book: *Geneticheskie Osnovy Mezhvidovoi i Mezhsortovoi Gibridizatsii Kartofelya i Pshenitsy.* Leningrad (Pushkin), Vol. 238, pp. 10–17.

9. Plekhanova, T.I. 1970. Prorastanie pyl'tsy kartofelya na pitatel'nykh sredakh (Germination of potato pollen on nutritional media). *Kartofel' i Ovoshchi*, No. 8, p. 11.
10. Matsubayshi, M.E. 1961. Cytogenic studies on Solanum Section Tuberarium with special reference to the interspecific relationships. Thesis, Hyogo Agricultural College, p. 121.
11. Oppenheimer, H.C. 1933. Cytogenetische Untersuchungen an Bastarde knollentragender *Solanum*-Arten. I. *Solanum chacoense* Bitt. × *Solanum tuberosum* L.s.str. Z.f. induktive Abstammungs lehre, 65, pp. 72–98.
12. Propach, H. 1938. Kreuzbarkeit von *Solanum*-Arten untereinander und mit Kulturkartoffeln und die Fertilität der Bastarde. *Forschungsdienst*, 6, pp. 273–280.
13. Swaminathan, M.S. and H.W. Howard. 1953. The cytology and genetics of the potato (*Solanum tuberosum*) and related species. *Bibliogr. Genetica*, 16, 192 pages.
14. Stelzner, G. 1943. Wege zur zuchterischen Nutzung des *S. chacoense* Bitt. in Hinblick, auf die Züchtung käferresistenter Kartoffelsorten. *Züchter*, 15, pp. 121–127.

TRUDY PO PRIKLADNOI BOTANIKE, GENETIKE I SELEKTSII

1978 Vol. 62, No. 1

Use of Wild Frost-Resistant Species of the Potato *S. commersonii* Dun. and *S. chomatophilum* Bitt. in Interspecific Hybridization

M.A. Vavilova
Candidate of Agric. Sciences

The wild diploid potato species S. commersonii and S. chomatophilum are resistant to frosts to −4, and −5°C and thus are of interest in breeding for frost resistance. They are almost uncrossable with varieties of S. tuberosum.

When S. commersonii and S. chomatophilum are crossed with tetra- and hexaploid species the problem of interspecific incompatibility is manifested, i.e., poor crossability, small number of seeds in the berry, low viability of the seeds and poor fertility of pollen in interspecific hybrids. The F_1 hybrids resist frosts to −3 and −5°C and may be taken as initial material for further breeding work.

Wild diploid potato species *S. commersonii* (series *Commersoniana* Buk.) and *S. chomatophilum* (series *Piurana* Hawkes), which can endure frosts from −4 to −5°C, are of considerable interest in potato breeding.

S. commersonii grows on the Atlantic coast of South America (Uruguay, East Argentina and Brazil) where it is represented by low-statured plants (height of the stem up to 20 cm), which have a stellar white and pale blue-purple corolla. Often the tubers of this species are formed under natural long-day conditions. Tubers are yellow and white, small, with a high content of dry matter; the stolons, in comparison with other wild species, are short or of average length (5–30 cm). Clones are found to be resistant to potato wart, common and powdery scab, viruses A and Y, stem nematode and Colorado potato beetle.

S. chomatophilum is found in Southern Colombia, Ecuador and Peru at an altitude of 2500–4000 meters above m.s.1. The plants are bushy, vigorous, up to 90 cm tall, with lustrous coarse leathery leaves. The corolla is wheel-shaped, large, and blue-purple (Fig. 1b). *S. chomatophilum* is related to species which hardly develop any tubers. Sometimes it does not develop stolons or tubers even under short-day conditions [1].

Fig. 1. The wild diploid species: a—*S. commersonii* Dun.; b—*S. chomatophilum* Bitt.

The above-named potato species have not been sufficiently used for hybridization in view of the difficulty in crossing them with species of other series.

Hybridization involving S. commersonii. R.W. Ross and P.R. Rowe [8] obtained hybrids from crosses of *S. commersonii* as the female parent with *S. megistacrolobum* Bitt., *S. multidissectum* Hawk., *S. sogarandinum* Ochoa, as well as from crosses involving *S. chomatophilum* as the female parent with *S. megistacrolobum*, *S. sanctae rosae* Hawk., *S. sogarandinum* and *S. vernei* Bitt. et Wittm.

Isolated hybrids were obtained from the hybridization of *S. commersonii* × *S. tuberosum* L. [2]. Improvement in crossability was achieved by using polyploids of *S. commersonii* (2n=48) in hybridization with tetraploid *S. tuberosum* and dihaploids of the varieties of *S. tuberosum* (2n=24) with diploid *S. commersonii* [3–5].

The results of our crosses of *S. commersonii* and *S. chomatophilum* with other wild and cultivated species of different series allow us to consider the possibility of their inclusion in hybridization (Table 1). Direct and reciprocal crosses of *S. commersonii* with varieties of *S. tuberosum* are not successful because of the genetic remoteness of these species and their different ploidy levels. When *S. commersonii* was used as the female parent the berries formed were mostly without seeds. Isolated seeds developed (on an average, 0.1 per pollinated flower) but these were not viable. In the hybridization of *S. commersonii* with the fertile dihaploid *S. tuberosum* we got a low percentage of successful crosses (7%). On the basis of the results of experimental studies we can confirm that in interspecific crosses *S. commersonii* is distinguished by its male sterility and, hence, it cannot be the male parent.

Table 1. Results of the crosses of S. commersonii Dun. and S. chomatophilum Bitt. and their hybrids with wild and cultivated potato species (1968–1977)

Combination	Number of pollinated flowers	Percentage of successful crosses	Quantity of seeds per berry	Quantity of seeds per pollinated flower
1	2	3	4	5
c × *S. vernei*	33	85	24	28
S. vernei × c	10	0	—	—
c × *S. megistacrolobum*	37	47	34	17
c × *S. sanctae rosae*	16	56	8	4
c × *S. boliviense*	9	44	1	0.4
c × *S. toralapanum*	13	46	17	8
c × *S. demissum*	43	56	28	15
c × *S. ehrenbergii*	8	25	1.5	0.4
c × *S. bulbocastanum*	31	0	—	—
c × *S. rybinii*	117	28	44	12
c × t 48	165	7	1.6	0.1
t 48 × c	40	0	—	—
c × t 24	82	7	39	2.9
t 24 × c	285	1	3	0.03
c × (*S. vernei* × *S. megistacrolobum*)	16	75	11	8
(c × *S. vernei*) × (*S. rybinii* × *S. vernei*)	7	0	—	—
(c × *S. vernei*) × t 48	17	0	—	—
(c × *S. megistacrolobum*) × (*S. rybinii* × *S. vernei*)	4	100	32	32

(*Contd.*)

1	2	3	4	5
(c × *S. megistacrolobum*) × *S. rybinii*	10	80	15	12
(c × *S. megistacrolobum*) × t 24	5	80	50	40
(c × *S. sanctae rosae*) × *S. rybinii*	36	3	2	0.06
(c × *S. sanctae rosae*) × t 48	68	0	—	—
t 24 × (c × *S. sanctae rosae*)	106	14	2.5	0.4
[t 24 × (c × *S. sanctae rosae*)] × t 48	12	0	—	—
(c × *S. toralapanum*) × *S. rybinii*	6	100	2.7	2.7
(c × *S. toralapanum*) × t 24	2	0	—	—
(c × *S. toralapanum*) × t 48	6	0	—	—
(c × *S. demissum*) × *S. rybinii*	9	67	1.3	0.9
(c × *S. demissum*) × t 48	19	21	1.5	0.3
t 48 × (c × *S. demissum*)	13	0	—	—
(c × *S. demissum*) × t 24	22	0	—	—
(*S. vernei* × *S. rybinii*) × c	75	1.3	1	0.01
(c × *S. rybinii*) × t 48	4	0	—	—
(c × *S. rybinii*) × t 24	14	0	—	—
[c × (t 24 × *S. toralapanum*)] × *S. rybinii*	2	100	33	33
[c × (t 24 × *S. toralapanum*)] × t 48	17	39	2	0.8
(t 24 × c) × t 48	11	9	3	0.3
c × ch	44	59	11	4
ch × c	15	0	—	—
(c × ch) × *S. vernei*	13	0	—	—
(c × ch) × *S. rybinii*	27	10	1	0.1
(c × ch) × (*S. rybinii* × *S. vernei*)	9	0	—	—
(c × ch) × *S. andigenum*	30	0	—	—
(c × ch) × t 48	8	0	—	—
(c × ch) × t 24	6	0	—	—
ch × *S. vernei*	10	10	12	12
S. vernei × ch	8	0	—	—
ch × *S. toralapanum*	5	0	—	—
ch × *S. sogarandinum*	14	29	40	11
S. verrucosum × ch	42	9.6	36	3.4
ch × *S. rybinii*	30	10	3	0.3
ch × t 24	18	6	1	0.06
t 24 × ch	64	1.6	2	0.03
ch × (*S. rybinii* × *S. vernei*)	17	23	10	2.3
(ch × *S. sogarandinum*) × t 24	8	0	—	—
(*S. verrucosum* × ch) × (*S. rybinii* × *S. vernei*)	9	11	11	1.2
(*S. verrucosum* × ch) × *S. rybinii*	38	24	1.8	0.4
(*S. verrucosum* × ch) × t 48	13	0	—	—
(*S. verrucosum* × ch) × t 24	20	0	—	—
t 24 × (*S. vernei* × ch)	25	0	—	—

Note: Legend followed here and afterwards: c—*S. commersonii*; t 48—variety of *S. tuberosum* (2n=48); t 24—diploid *S. tuberosum* (2n=24); ch—*S. chomatophilum*.

Crosses with *S. commersonii* as the female parent were successfully conducted with some diploid species: *S. vernei*, *S. boliviense* Dun., *S. toralapanum* Card. et Hawk., *S. megistacrolobum*, *S. sanctae rosae*, *S. chomatophilum*, *S. rybinii* Juz. et Buk. and with hexaploid *S. demissum* Lindl. Here the most successful combinations of *S. commersonii* were with the following species taken as the male parent: *S. vernei*, *S. megistacrolobum*, *S. demissum* and *S. rybinii* with an average of 24–44 seeds in a berry. As the wild Mexican diploid species are distinguished by genetic incompatibility in interspecific crosses, the crosses *S. commersonii* × *S. bulbocastanum* Dun. were unsuccessful. In the combination *S. commersonii* × *S. ehrenbergii* (Bitt.) Rydb. we got three seeds from eight pollinated flowers.

The characteristic peculiarities of hybrid seeds obtained through the involvement of *S. commersonii* are small size, poor germinative vigor and low viability. For example, the size of the seeds of *S. commersonii* is 2 × 1.4 mm, but in crosses of this species with *S. chomatophilum* it is 1.1 × 0.9, with *S. vernei* —1 × 0.7, with *S. toralapanum*—0.8 × 0.7, with *S. megistacrolobum*—1.5 × 1.0, with *S. rybinii*—1.2 × 0.9 and with *S. demissum*—2 × 1.3 mm. In this way, the seeds of the hybrids are smaller than those of the female parent. Germination of hybrid seeds began on the 7th–10th day and continued for a period of 60–70 days, i.e. the germinative vigor was very low. Viability of seeds from crosses of *S. commersonii* with *S. vernei* was 13%, with *S. boliviense*—25%, with *S. toralapanum*—5%, with *S. megistacrolobum*—4%, with *S. sanctae rosae*—8%, with *S. chomatophilum*—5%, with *S. demissum*—8% and with *S. rybinii*—11%.

Anomalies were observed in the germination of seeds and the growth of saplings. In the populations of *S. commersonii* × *S. rybinii* and *S. commersonii* × *S. vernei* the sprouts of the seeds were thickened and they developed roots similar to sprouts cultivated in colchicine; such germinated seeds, when sown in the soil, did not give seedlings. In certain saplings we noticed ring like thickened cotyledons; even tricotyledonous saplings were observed which died during the cotyledon stage. The low viability of seeds could have been explained by their small size and, hence, paucity of nutritional substances in the endosperm required for the development of the sprouts. However, low viability of seeds was observed in the combination *S. commersonii* × *S. demissum*, also, whose seeds were considerably bigger (2 × 1.3 mm). Low viability of seeds, their small size and the deformity of sprouts are problems of incompatibility, which are inherent in *S. commersonii* in interspecific crosses. Moreover, the problem of incompatibility manifested itself in the low fertility of the pollen of interspecific hybrids. For example, the quantity of colored pollen grains in combinations of *S. commersonii* with different species was as follows: with *S. vernei*—29%, with *S. rybinii*—26%, with *S. toralapanum*—38%, with *S. sanctae rosae*—44%, with *S. megistacrolobum*—34% and with *S. demissum*—74%.

Self-incompatibility of hybrids was also often observed. It was only in three combinations that we could get isolated berries through forced self-pollination (*S. commersonii* × *S. megistacrolobum*, *S. commersonii* × *S. rybinii*, *S. commersonii* × *S. demissum*) which enabled us to use these hybrids in subsequent crossings mostly as the female parent. However, these crosses were not as successful (low percentage of berry formation; the berries, if formed, had no seeds or had only a few seeds). The following crosses were more successful: [(*S. commersonii* × *S. megistacrolobum*) × *S. tuberosum* (2n = 24)], [(*S. commersonii* × *S. megistacrolobum*) × *S. rybinii*], [*S. commersonii* × *S. megistacrolobum*) × (*S. rybinii* × *S. vernei*)] and [*S. commersonii* × *S. tuberosum* (2n = 24) × *S. toralapanum*] × *S. rybinii*.

Crossings involving S. chomatophilum. Despite the fact that this species is distinguished by its high frost resistance and resistance to Phytophthora it is hardly used in breeding work either in our country or abroad. According to our observations it can be used in crossings both as the female as well as the male parent.

S. chomatophilum is not easily crossed with other species: when it was used as the female parent in crosses with *S. toralapanum*, with *S. spegazzinii* as well as in the combination *S. vernei* × *S. chomatophilum* berries did not form. However, crosses of *S. chomatophilum* with the following species were made: *S. rybinii*, *S. vernei*, *S. sogarandinum* and also *S. commersonii* × *S. chomatophilum* and *S. verrucosum* × *S. chomatophilum*. For example, in the combination *S. chomatophilum* × *S. vernei* 10% of the crossings were successful; the average number of seeds in the berry was 12 and per pollinated flower 1.2. The most successful was the combination *S. chomatophilum* × *S. sogarandinum* (where the average number of seeds was 31).

Subsequent crossings showed that hybrids of *S. chomatophilum* are not easily crossed with diploid and tetraploid components. Hybrids of *S. commersonii* × *S. chomatophilum* are self-incompatible and do not form berries through forced self-pollination. Apart from that, they were pollinated with pollen from *S. vernei*, *S. rybinii* and *S. andigenum*, but seeds were not formed. When pollinated with pollen from *S. rybinii* and *S. andigenum* parthenocarpic berries developed from each of the pollinated flowers. The efforts to cross the hybrids with varieties and dihaploids of *S. tuberosum* proved without result.

Hybrids of *S. verrucosum* × *S. chomatophilum* are partially self-compatible and fertile; 33% of the pollen grains were colored. Berries developed through self-pollination; the number of seeds in a berry was 1–68. Efforts to cross the hybrids with *S. tuberosum* (2n = 24 and 48) were without result. When the hybrids were pollinated with pollen from *S. rybinii*, isolated berries with a few seeds were obtained (the average number of seeds per berry was 1.8 and per pollinated flower 0.4). Pollination of the hybrids of *S. verrucosum* × *S. chomatophilum* with pollen from the hybrids of *S. rybinii* × *S. vernei* resulted in an improvement in crossability (the average number of seeds in a berry was 11 and per pollinated flower 1.2).

Characteristics of the F_1 hybrids. In the hybrids of *S. commersonii* the following characters were dominant under natural day length: round shaped and blue-purple color of the corolla, long stolons and poor formation of tubers. Only in the combinations *S. commersonii* × *S. vernei*, *S. commersonii* × *S. chomatophilum* and *S. commersonii* × *S. rybinii* were blue flowers formed on the hybrids (Fig. 2); individual plants were also observed with white flowers. The heterogeneity of the color of the corolla was characteristic of hybrids of *S. chomatophilum* × *S. sogarandinum*. Flowering was profuse, and parthenocarpic berries developed by the end of vegetative growth in hybrids of this combination.

Fig. 2. A plant of the hybrid 41/77 (*S. commersonii* × *S. rybinii*).

The characters of the male parent dominate in the formation of the corolla in the hybrids of *S. commersonii* × *S. demissum*. The external side of the corolla is more intensely colored than the internal. Hybrids are self-compatible and form berries through natural self-pollination (fertility of pollen, 74%), but they are difficult to cross. Only a few seeds are obtained from the crosses of *S. rybinii* with varieties of *S. tuberosum*.

One of the most significant shortcomings of the hybrids of *S. commersonii* and *S. chomatophilum* is the absence of tuber formation under natural day length. It was only at the end of October under short-day conditions in a greenhouse that isolated small tubers measuring 0.7–1.5 cm in diameter were formed in the crosses of *S. commersonii* × *S. vernei*, *S. commersonii* × *S. megistacrolobum* and *S. tuberosum* (2n=24) × *S. commersonii*; at the beginning of December tubers developed in the combinations *S. commersonii* × *S.*

Table 2. The characteristics of interspecific hybrids involving S. commersonii Dun. and S. chomatophilum Bitt.

Origin of hybrids	Plant height, cm	Color of corolla	Frost resistance estimated in points			Formation of	
			−3°C	−4°C	−5°C	berries	tubers
c × *S. vernei*	15–30	white, Bp*	9	6–7	5–6	—	+
c × ch	45–50	white, Bp	9	7–8	7–8	—	—
c × *S. megistacrolobum*	20–25	Bp	9	7–9	6–9	+	+
c × *S. sanctae rosae*	30–35	,,	9	9	7	—	—
c × *S. boliviense*	30–40	,,	9	9	7	—	+
c × *S. toralapanum*	10–15	,,	9	7–8	7	—	—
c × *S. demissum*	20–25	,,	9	7–9	7–9	+	+
c × *S. rybinii*	25–30	white, Bp	7	5–7	5	+	—
S. tuberosum 2n=24 × c	30–40	white	8	7–8	—	—	+
c × (*S. tuberosum* × *S. megistacrolobum*)	25–35	Bp	9	6–7	6	—	—
c × (*S. tuberosum* × *S. toralapanum*)	25–35	,,	9	6–7	6	—	—
ch × *S. vernei*	35–40	,,	9	7–9	5–7	—	—
ch × *S. sogarandinum*	15–25	,,	9	9	9	—	—
ch × *S. rybinii*	40–50	,,	9	7	5–7	—	—
S. verrucosum × ch	30–40	,,	8	5–8	6	+	+
S. commersonii	15–25	white	9	9	7–8	+	+
S. chomatophilum	40–55	Bp	9	7–9	7–9	+	—

Note: Scale (points) of resistance: 9—no damage; 7—young leaves damaged (up to 20%); 5—50% of leaves and apex of stems frost damaged; 3—lower leaves and bases of the stems remained undamaged; 1—complete death of plants.

*Bp—blue-purple.

demissum, *S. verrucosum* × *S. chomatophilum* (Table 2). Hybrids of the cross *S. verrucosum* × *S. chomatophilum* were also interesting. According to reports of G.E. Marks [7], hybrid plants of *S. verrucosum* × *S. chomatophilum* were not viable and they died while using their own roots as well as when grafted on tomatoes. In our experiments the morbidity of the hybrids manifested itself in the growing phase of sprouts and cotyledons. Plants of the saplings were small, with chlorophyll deficiency and blue-purple anthocyanin. Out of 50 germinated seeds, 39 gave seedlings when the seeds were sown, but only 22 saplings among them survived. In contrast to foreign data, in our experiments the plants developed very well subsequently. As regards the bush habit, the hybrids were similar to *S. verrucosum* (Fig. 3). Leaf lobes had a short pubescence on the ventral side (leaves of *S. chomatophilum* are rough, not pubescent). The corolla was round, and blue-purple, with the dorsal side more intensely colored as is inherent in *S. verrucosum*. The hybrids are fertile, with well developed berries through natural self-pollination. The surface of the berries is smooth as in *S. chomatophilum*.

Fig. 3. A plant of the hybrid 51/77 (*S. verrucosum* × *S. chomatophilum*).

Cytological analysis done on the slides prepared from the root tips showed that all the hybrids under study obtained from crosses of the diploid species with each other were found to be diploids (2n=24). We tried to transform individual hybrids of *S. chomatophilum* to the tetraploid level with the idea of using them in crossings with *S. tuberosum*. For this purpose we put 0.5% aqueous solution of colchicine on the growing point of the hybrids of *S. commersonii* × *S. chomatophilum*, *S. chomatophilum* × *S. rybinii*, *S. chomatophilum* × *S. vernei*, *S. chomatophilum* × *S. sogarandinum* and saplings of *S. chomatophilum*. However, all the treated hybrids were found to be diploid. Probably, a much higher concentration of colchicine is required to induce polyploidy in *S. chomatophilum* and its hybrids.

Certain combinations are of considerable interest regarding resistance to Phytophthora. For example, hybrids from the crosses *S. verrucosum* × *S. chomatophilum* and *S. commersonii* × *S. demissum* were not infected by this disease under field conditions.

Frost resistance of hybrids of S. commersonii and S. chomatophilum. Study of frost resistance was conducted at temperatures of −3, −4, −5°C for a period of four hours. Hybrids withstood a frost of −3°C, well; however, in populations which were developed from crosses involving one susceptible parent component (*S. commersonii* × *S. rybinii*, *S. tuberosum* 2n=24 × *S. commersonii*, *S. verrucosum* × *S. chomatophilum*) we observed frost damage to the upper leaves in certain hybrids (Table 2).

The effect of a temperature of −4°C led to a more pronounced manifestation of the differences in frost resistance between hybrids of different origin. Most resistant, without the slightest damage, were *S. commersonii* and hybrids of *S. commersonii* × *S. boliviense*, *S. commersonii*, *S. sanctae rosae*, *S. chomatophilum* × *S. sogarandinum*.

Slight damage to young leaves (up to 20%) was observed in the hybrids *S. commersonii* × *S. toralapanum*, *S. commersonii* × *S. chomatophilum*, *S. commersonii* × *S. megistacrolobum*, *S. commersonii* × *S. demissum* and *S. chomatophilum* × *S. vernei*. Hybrids, whose genealogy included resistant as well as less resistant (*S. vernei*), and frost susceptible species (*S. rybinii*, *S. tuberosum*, *S. verrucosum*), were considerably damaged (up to 40–50% of the leaves). Most resistant were the hybrids from crosses of two frost-resistant species, *S. chomatophilum* × *S. sogarandinum*, which endured a frost from −3 to −5°C.

The data from our studies provide a basis to assume that the wild diploid species *S. commersonii* and *S. chomatophilum* are of interest as initial material in breeding for resistance to frosts.

Improvements in the economically important traits of hybrids can be achieved through their advance polyploidization and subsequent crossings with varieties of *S. tuberosum*.

CONCLUSIONS

1. Wild diploid species *S. commersonii* and *S. chomatophilum* tolerate frosts to −4 and −5°C and they can be used as initial material in breeding for frost resistance.

2. In the hybridization of this potato species with wild, cultivated and diploid species there are problems of poor crossability, small quantity of seeds in the berry, their low viability, sterility or poor fertility of the pollen of interspecific hybrids. Both species rarely can be crossed with *S. tuberosum.*

3. Frost resistance, long stolons and poor tuber formation are dominant in the F_1 hybrids. Hybrids are resistant to frosts from −3 to −5°C.

4. In order to get varieties with economically important traits it is necessary to conduct recurrent crossings of hybrids with specially selected varieties of *S. tuberosum.*

REFERENCES

1. Bukasov, S.M. 1971. Dikie vidy kartofelya (Wild species of the potato). In the book: *Kul'turnaya Flora SSSR. Kartofel'*, Vol. 9, pp. 5–40. Leningrad.
2. Bukasov, S.M. and A.Ya. Kameraz. 1972. *Selektsiya i Semenovodstvo Kartofela* (Breeding and seed technology of the potato). Leningrad.
3. Laptev, Yu.P. and I.G. Nabokov. 1970. Ispol'zovanie mutatsii khromosomnogo kompleksa dlya preodoleniya mezhvidovoi nesovmestimosti kartofelya (Use of mutations of the chromosome complex to remove interspecific incompatibility of the potato). *Genetika*, Vol. 5, No. 11, pp. 30–40.
4. Gorea, T. 1970. Fertilitat and Kreuzbarkeit der Dihaploiden von *Solanum tuberosum* L. und deren F_1 Bastarden. Zeitschrift für Pflanzenzüchtung, Vol. 64, No. 3, pp. 201–220.
5. Hougas, R.W. and S.J. Peloquin. 1958. The potential of potato haploids in breeding and genetic research. *Am. Potato J.*, Vol. 35, No. 10, pp. 701–707.
6. Irikura, Y. 1968. Studies on interspecific crosses of tuber-bearing Solanums. I. Overcoming cross incompatibility between *S. tuberosum* and other *S.* species by means of induced polyploids and haploids. Reprint from the Research Bull., No. 92, pp. 21–38.
7. Marks, G.E. 1965. Cytogenetic studies in tuberosum Solanum species. III. Species relationships in some South and Central American species. *The New Phytologist*, Vol. 64, No. 2, pp. 293–306.
8. Ross, R.W. and P.R. Rowe. 1969. Utilizing the frost resistance of diploid *Solanum* species. *Am. Potato J.*, Vol. 46, No. 1, pp. 5–13.

TRUDY PO PRIKLADNOI BOTANIKE, GENETIKE I SELEKTSII

1978 Vol. 62, No. 1

Resistance to Phytophthora Infection in Hybrids Obtained from Crosses of the Dihaploid *S. tuberosum* L. with Wild and Cultivated Forms of *Solanum*

K.Z. Budin
Academician of the VASKhNIL

V.L. Broksh, M.M. Khromtsova
Candidates of Biol. Sciences

Hybrids resulting from the crossing of dihaploids of Solanum tuberosum with wild and cultivated species of Solanum have been tested for their resistance to Phytophthora infestans. The tests were carried out on the haulms and tubers by laboratory and field methods. The transmission of field resistance to Phytophthora infestans was found to be possible at the diploid level in F_2*. In this connection prospects are outlined for producing interspecific hybrids with far better resistance to the pathogen than that in the parental forms.*

Among the more than 20 fungal diseases of the potato, the late blight of potato [pathogen—*Phytophthora infestans* (Mont.) de Bary] remains the most harmful. In the development of varieties highly resistant to this disease, it is of great importance to study the peculiarities of the inheritance of the resistance character.

Resistance of potatoes, to Phytophthora manifests itself as a result of the interaction of two genetic systems. Two types of resistance are distinguished:

polygenic (field) resistance to all existing races of Phytophthora and monogenic or "Super sensitivity" to individual races of the pathogen; resistance to the races is controlled by dominant genes.

Field resistance is of a complex nature and is inherited polygenically; therefore it forms in the progeny of susceptible parents. Moreover, it has been observed that the average resistance of the progeny is correlated with the resistance of the parents. Field resistance is characteristic in varieties originating from the wild species. For this purpose *Solanum demissum* Lindl. and *S. stoloniferum* Schlechtd. are used most extensively [1]. Other wild species which are distinguished by the nature and specificity of the action of the genes have also been studied during the past few years and their application to a breeding program can enable us to solve the problem of long-lasting resistance to Phytophthora. However, crossing cultivated species with them, particularly with the group of diploid species, is very difficult. There are two ways to improve crossability: advance or preliminary polyploidization of species or dihaploidization of varieties. The second way is preferable because of the possibility of conducting selection among the hybrids at the diploid level. It is known that diploids of the same variety vary considerably regarding resistance to Phytophthora [5]. Here, the more resistant the initial variety, the greater will be the probability of getting resistant dihaploids. In this way dihaploid forms with high resistance were found in 1967 [4]. In 1971–1974 we got relatively resistant dihaploids and hybrids from crosses with them.

The studies were conducted at the Moscow branch of the VIR-All Union Research Institute of Plant Industry (Mikhnevo—Moscow province). We included 20 dihaploids of the following varieties in our studies: Amsel, Antje, Apta, Everest, Priekul'skii rannii, Saco, Sebago and hybrids of these dihaploids with species *S. phureja* Jug. et Buk., *S. kesselbrenneri* Juz. et Buk., *S. vernei* Bitt. et Wittm., *S. demissum* Lindl., *S. chacoense* Bitt., *S. raphanifolium* Card. and a group of complex hybrids from crosses of dihaploids of *S. andigenum* Juz. et Buk. with the species *S. stenotomum* Juz. et Buk. and *S. pampasense* Hawk.

The plants were grown in the collection nursery in ten-tuber plots in three replications. In the families from crosses of dihaploids with the diploid species the number of hybrids varied from 16 to 68 per family; while developing the complex hybrids the number of families varied from 80 to 324 hybrids, whereas in the crossings of sibs[1] it varied from 10 to 150. In some families in the latter case we could not get any progeny.

Phytophthora resistance was estimated by field and laboratory methods. Field resistance of the aerial parts of plants was evaluated during natural development of the disease as per the five-point system of the VIR (0—absence of infection, 5—maximum points) and during artificial infection of leaves with a

[1]Sibs—sister crossings.

Table 1. Phytophthora infection of dihaploids and initial tetraploid varieties (estimated in points) under field conditions (1971–1974)

Varieties and dihaploids	Ploidy level (2n)	Haulms			Tubers		
		1971	1973	1974	1971	1973	1974
Everest	48	0	0	0	0	0	0
Everest 20/63	24	3	5	0	0	1	0
„ 21/63	24	1	4	0	0	0	0
„ 23/63	24	4	4	1	1	0	0
„ 31/63	24	0	0	0	1	0	0
Amsel	48	4	1	1	0	2	1
Amsel 258	24	4	5	1	0	2	1
„ 259	24	4	4	1	3	3	1
„ 262	24	5	5	1	0	3	1
„ 263	24	4	5	1	0	2	1
Apta	48	0	2	0	0	1	0
Apta 271	24	4	5	1	1	2	1
„ 272	24	2	4	0	0	1	0
„ 273	24	4	4	1	0	1	0
Antje	48	5	5	1	1	1	0
Antje 2	24	5	5	1	0	3	1
„ 6	24	3	5	1	0	3	1
Priekul'skii rannii	48	5	5	2	2	4	1
Priekul'skii rannii 1	24	4	4	1	1	3	1
Priekul'skii rannii 2	24	3	3	0	0	2	0
Saco	48	2	4	1	0	3	0
Saco 636	24	2	4	1	0	1	0
„ 649	24	5	5	1	0	0	0
„ 644	24	2	4	3	0	1	1

culture of highly virulent races 1.2.3.4. The degree of resistance of the tubers was determined according to the method of Mooi [3], by infecting all the tubers with a culture of race 1.

The degree of Phytophthora infection in dihaploids and initial varieties varied under field conditions from year to year during our studies. The maximum development of the disease on the haulms as well as on the tubers was observed in 1973. Results of the estimation of field resistance to Phytophthora of 18 dihaploids of the potato and their initial varieties (Table 1) showed a positive correlation between infection of haulms and tubers (r=0.6 and 0.5 in 1971 and 1974, respectively). According to the 1973 data, the correlation was insignificant and not reliable (r=0.1). Under field conditions the dihaploids had infection either at par with the initial varieties or somewhat higher. An exception was one dihaploid of the variety Priekul'skii rannii which, according to the data on field resistance as well as laboratory observations, was infected less than the initial form. We had two dihaploids of this variety and in both of them there was no sporulation when leaves were infected by race 0, whereas in the case of infection of the initial form, we noticed profuse sporulation.

Artificial infection of the leaves by the most virulent race 1.2.3.4 did not differ significantly between the resistance of the dihaploids and tetraploid varieties. Hybrids of dihaploids with diploid species (second vegetative reproduction), when infected with the race 1.2.3.4, differed among themselves regarding their resistance even in cases where the parents showed similar resistance. Dihaploids of variety Apta gave a different spectrum of susceptibility in the progeny from crossings with the same species (Table 2). In the dihaploid of Apta 272 we noticed more plants with slight infection in the

Table 2. Results of the infection of Phytophthora race 1.2.3.4, in the leaves of hybrids from crosses of dihaploids of S. tuberosum with wild and cultivated species of Solanum

Combination	Infection of parents, in points		Percentage of infected hybrids	
	♀	♂	up to 2 points	3–5 points
Apta 272 × *S. kesselbrenneri*	3	3	74	25
„ 271 × *S. kesselbrenneri*	3	3	16	82
„ 272 × *S. vernei*	3	3	100	—
„ 271 × *S. vernei*	3	3	64	36
„ 272 × *S. phureja*	3	4	99	1
Sebago 808 × *S. kesselbrenneri*	3	3	45	55
„ 808 × *S. demissum*	3	0	95	5
Amsel 259 × *S. chacoense*	3	4	49	51

Table 3. Results of the infection of leaves of complex diploid hybrids by the race 1.2.3.4

No. of families	Combination	Infection of parents, in points		Percentage of infected hybrids	
		♀	♂	up to 2 points	3–5 points
	Sibs				
1	(Apta 272 × *S. kesselbrenneri*) × (Apta 272 × *S. kesselbrenneri*)	1	2	100	—
2	To same	1	3	80	20
3	,,	3	2	20	80
1	(Sebago 808 × *S. kesselbrenneri*) × (Sebago 808 × *S. kesselbrenneri*)	2	2	83	17
	Backcrosses				
1	(Apta 272 × *S. kesselbrenneri*) × *S. chacoense*	2	4	48	52
1	(Apta 272 × *S. kesselbrenneri*) × *S. raphanifolium*	2	3	28	72
1	(Apta 272 × *S. kesselbrenneri*) × *S. vernei*	2	3	23	77
2	To same	1	3	61	39

progeny than in Apta 271. Even in crosses with the highly susceptible form of *S. phureja* of the dihaploid of Apta 272 had a low infection rate.

The resistance of the progeny somewhat increased in crossings with the involvement of *S. vernei* as the male parent which has low resistance. In crosses with *S. demissum* all hybrids had infection to some degree, and fully immune saplings were not noticed. As a whole, the majority of hybrid plants excelled their dihexaploid parents regarding resistance. Plants more susceptible than the parent forms were found only in the families: Apta 271 × *S. kesselbrenneri*, Sebago 808 × *S. kesselbrenneri*, Amsel 259 × *S. chacoense*.

The resistance of progeny from crosses of sibs varied depending upon the resistance of the forms involved in the crossings. In three out of four families, the majority of saplings were slightly infected by the race 1.2.3.4 (Table 3).

While affecting artificial infection of the tubers with Phytophthora (race 1) it was established that among dihaploids from varieties with highly susceptible tubers only one (Everest 21/63) had a slight infection. The remaining ones were highly susceptible. Dihaploids of the varieties Apta and Saco were distinguished by the relative resistance of the tubers.

In trials involving simple hybrids from crosses with species *S. kesselbrenneri*, *S. phureja* and *S. chacoense*, up to 60% of the saplings had tubers with a slight infection. In these crosses only a few isolated plants had tubers without infection.

Hybrids from crosses involving species *S. vernei* and *S. demissum* were distinguished by greater resistance. For example, in the family Sebago 808 × *S. demissum*, for out of 16 hybrids had tubers completely free from infection while the remaining ones had infection not exceeding 1 point.

In the crosses of dihaploids of Apta 272 × *S. vernei* there was no infection in tubers in 11 out of 54 hybrids; only in four cases was the infection considerable (3–4 points). In the rest of the hybrids the degree of infection of tubers did not exceed 1–2 points. In the family of the hybrids of Apta 271 × *S. vernei*, 13 out of 49 saplings did not have infected tubers; however, highly infected tubers (3–5 points) were found in five families.

Crosses of the hybrids so obtained with other wild species did not lead to an increase in the resistance of the progeny: the majority of the saplings were infected to a great extent, except in the second family (Apta 272 × *S. kesselbrenneri*) × *S. vernei* (Table 3).

Field observations in different years showed that families were infected by Phytophthora differently, but there were no significant differences in resistance between families from crosses with the species *S. chacoense* and *S. raphanifolium*. Only in crosses with *S. vernei* were relatively resistant hybrids in the majority in the progeny. Similarly, resistant forms were obtained when a resistant hybrid from the family of Apta 272 × *S. kesselbrenneri* was used as the female parent in the cross (Tables 3 and 4).

Sporulation of the pathogen was observed only on individual tubers in the families: Apta 271 × *S. kesselbrenneri* and Sebago 808 × *S. kesselbrenneri*.

When there is infection in the tubers of saplings from the crosses of sibs, sporulation of race 1 was observed in individual plants in all variants of the crossings. Simultaneously, forms without any infection in the tubers were also isolated. For example, out of 64 hybrids of the family Sebago 808 × *S. kesselbrenneri* five hybrids had tubers without Phytophthora infection while the infection in tubers of the remaining saplings varied from 1 to 4 points.

In complex hybrids the forms which had tubers free from Phytophthora infection were isolated in the combinations: (Apta 272 × *S. kesselbrenneri*) × *S. chacoense* and in both families of (Apta 272 × *S. kesselbrenneri*) × *S. vernei* (Table 4).

Table 4. Results of artificial infection by Phytophthora (race 1) of tubers from hybrids with different degrees of resistance in the haulms under field conditions

No. of families	Combination	Plants with infected haulms, %			Plants with resistant tubers, %
		1973	1974	1975	
1	(Apta 272 × *S. kesselbrenneri*) × *S. chacoense*	95	75	0	1
1	(Apta 272 × *S. kesselbrenneri*) × *S. raphanifolium*	30	50	10	0
1	(Apta 272 × *S. kesselbrenneri*) × *S. vernei*	50	18	3	1
2	To same	15	2	0	19

As we know, there is no direct correlation between resistance of leaves and tubers to Phytophthora. Forms combining both these traits are quite rare. We could isolate only four hybrids with resistant haulms and tubers: two from the crosses of sibs of the family of Sebago 808 and *S. kesselbrenneri*, one from the combination (Apta 272 × *S. kesselbrenneri*) × *S. vernei* and one from the cross (Apta 272 × *S. kesselbrenneri*) × *S. chacoense*.

Hybrids of dihaploids of *S. andigenum* were studied from 1971 to 1974. As a result, four forms, viz. 240, 242, 244 and 245 with high field resistance to Phytophthora were isolated from the family (*S. andigenum* 2787 × *S. stenotomum* 140) × *S. pampasense*. One of the sibs (240) had high male fertility and was used as the male parent in crosses with forms having high field resistance—sibs 242, average field resistance—hybrid 46 (Amsel 259 × *S. chacoense*) and low field resistance—*S. chacoense* (from the crosses of two lines). In the progeny from crosses of the forms with high field resistance (sibs 242 × 240) the majority of hybrids had average resistance; resistant hybrids were 17% and susceptible forms were 16% (Table 5).

Table 5. Results of infection by the race 1.2.3.4 of leaves of hybrids from the crosses of diploid forms of potato with phytophthora-resistant hybrid-dihaploid 240

Combination	Degree of resistance	Percentage of hybrids		
		susceptible	average resistance	resistant
242 × 240	Resistant	16	67	17
46 (Amsel 259 × *S. chacoense*) × 240	Average resistance	86	14	0
F_1 *S. chacoense* × 240	Susceptible	84	16	0
F_2 *S. chacoense* × 240	,,	56	22	22

Resistant saplings were not found in families from crossings of average resistance and susceptible forms with the hybrid 240. We could not get an F_2 from the cross 242 × 240. In the F_2 of family 46 × 240 we got only 10 viable plants; death of the seedlings was mainly due to the emergence of albinos (35%). A more viable second generation was obtained in the family *S. chacoense* × 240. Among the seedlings there were only 5% albinos. When leaves of the F_2 of this family were artificially infected by the race 1.2.3.4 we found resistant forms, although they were not observed in the F_1.

As a result of the study of field resistance to Phytophthora of dihaploids and initial tetraploid varieties, we could not detect any significant differences between them. The only exceptions were two dihaploids from the variety Priekul'skii rannii.

The incidence of positive inheritance of field resistance to Phytophthora observed in progeny from self-pollination of the potato, is evidently possible during the formation of dihaploids as well. From the reports available in literature it is known that an increase in the level of field resistance of hybrids can take place even when parent forms themselves have only a low degree of resistance [2]. Field resistance is controlled by multiple genes; therefore, the percentage of progeny with a much higher level of resistance depends on the possibility of genes of the parents supplementing one another. It is through this mutual complementary factor of resistance of dihaploids and diploid species that we can explain the emergence of hybrids with field resistance in our experiments in crosses with species *S. phureja*, *S. kesselbrenneri* and *S. chacoense*, which do not have high resistance. This explanation is also confirmed by the appearance of resistant hybrids in the second generation, and their absence in the F_1 in the family obtained from crosses of susceptible forms with resistant ones.

Recurrent crossings of the hybrids so obtained with some wild species did not lead to an increase in the number of resistant forms in the progeny when compared with the primary crossing (dihaploid × species). An increase in the average resistance of saplings was observed only when a more resistant hybrid was selected as the female parent. In saplings from such crosses there is a deterioration in such traits as size and shape of tubers and length of the stolons.

Resistance of the hybrid progeny is determined to a great extent by the resistance of the parent forms. Even in dihaploids of the same variety, with almost the same resistance, a different spectrum of variability of this trait occurred in the progeny. However, when crossing hybrid 240 with the plasma of *S. andigenum* as the male parent there was no significant difference between families from its crosses with susceptible and average resistant forms.

The second generation of diploid hybrids obtained through crosses of sibs, irrespective of which dihaploids (*S. tuberosum* or *S. andigenum*) were used in the crosses, had in the majority of cases a much higher field resistance than in the F_1. Two of the four hybrids with high field resistance to race 1 of Phytophthora (of haulms and tubers) were obtained from such crosses.

All the isolated forms have short stolons and large tubers, typical of the cultivated potato. In our experiments we did not find saplings with traits predominantly of the wild parent in the second generation. This was probably due to the fact that at the diploid level there occurs a much more complete regrouping of the chromosomes during gamete formation than at the tetraploid level. In this process there is conjugation of the homologous chromosomes of the parents and not of sister chromosomes of one species as may occur in selective conjugation at the tetraploid level. This capability of more complete recombination of the genes of the parent forms of diploid hybrids can be one of the several advantages of preliminary breeding at the diploid level.

CONCLUSIONS

1. In the potato it is possible to have cases of inheritance of field resistance to Phytophthora at the diploid level while developing dihaploids, as well as at the level of simple hybrids involving diploid species, particularly in the second generation from crosses of the sibs.
2. The average level of field resistance of the diploid hybrids is determined by the degree of resistance of the parents.
3. Breeding the potato at the diploid level opens up new prospects for getting interspecific hybrids with increased resistance to Phytophthora.

REFERENCES

1. Kotova, K.A. 1976. Novyi istochnik ustoichivosti k fitoftore (A new source of resistance to Phytophthora). *Selektsiya i semenovodstvo*, No. 1, p. 30.

2. Black, W. 1970. The nature and inheritance of field resistance to late blight (Phytophthora infestans) in potatoes. *Am. Pot. J.*, 8, p. 279.
3. Mooi, I. Metoden zur Festställung der Resistenz der Knollen von Kartoffelsorten, gegenuber Phyt. infestans in Laboratorium. Simpos. Problem der Resistenz gegenüber Phyt.
4. Peres-Ugalge, P. and S.L. Peloquin. 1964. Potato haploid with field resistance to late blight. *Am. Pot. J.*, 44, p. 9.
5. Sosa Chaver, R. and M. Hernandez de Sosa. 1970. The use of dihaploids in breeding *Solanum tuberosum* L. in Mexico. *Hereditas*, 64, 2, p. 243.

TRUDY PO PRIKLADNOI BOTANIKE, GENETIKE I SELEKTSII

1978 Vol. 62, No. 1

Development and Use of Inbred Lines of the Potato in Inter-line and Varietal-line Hybridization

K.Z. Budin
Academician VASKhNIL

T.I. Soboleva
Candidate of Agric. Sciences

Inbred lines I_1–I_3 of interspecific potato hybrids were obtained as a result of crossing Solanum andigenum with diploid primitive species. The latter had been singled out for its set of beneficial properties. These lines were used in crosses among themselves as well as with tetraploid varieties. Among their progeny, economically valuable hybrids were selected, which had an advantage over regionally adapted varieties. This paper emphasizes the need to create inbred lines of interspecific hybrids to obtain valuable forms for varietal-line and inter-line hybridization.

The development and use of inbred lines is one of the fundamental functional links in the breeding of many agricultural crops. Self-pollination carried out over several generations increases the degree of homozygosis of the material, which simplifies subsequent work on the development of forms with a complex of desirable characters, so that the entire latent genetic information of the population is revealed and the selection of desirable genotypes becomes easy.

In a number of crops (corn, beet and others) inter-line and varietal-line hybrids proved to be particularly effective.

In breeding the potato, a crop with high heterosis, it is particularly important to increase the degree of homozygosis in the components of crossings. However, it is not always possible to get homozygous lines in the varieties because of the absence of self-pollination or because of drastic depression of the inbred lines. Krantz, a scientist from the USA [5–7], came to the conclusion that more care should be taken in the use of the first and second inbred lines because prolonged self-pollination affects the yield adversely, reducing it by 40% in comparison with the F_1. He reported an advantage in the yield of varietal-line hybrids, not only over intervarietal hybrids, but also over hybrids of the type Variety $\times F_1$.

In our country the method of inter-line and varietal-line hybridization is not receiving much attention. Insignificant research in this direction is being conducted at the Byellorussian Research Institute of Potato, Fruit and Vegetable Cultivation (BNIIKPO), and at the Institute of Potato Farming (Moscow). There the scientists have developed inbred lines of hybrid origin with high starch content and *Phytophthora* resistance.

In the department of genetics at the VIR (All-Union Research Institute of Plant Industry) work has been conducted over a number of years (1971–1978) on the development and genetic study of self-pollinated lines of interspecific hybrids in comparison with double-triple specific hybrid forms. Lines were obtained from interspecific hybrids developed by us through the crossing of dihaploids of *Solanum andigenum* × diploid species (*S. stenotomum* or *S. rybinii*) with subsequent spontaneous doubling of the chromosome number in the F_2. From some highly fertile hybrid forms of the F_2 we obtained lines which were equal to the initial hybrids in regard to yield and the quality of the tubers. It is evident from Table 1 that among the small number of clones in the first hybrid combination (dihaploid *S. andigenum* × *S. stenotomum*) the F_2 and I_1 are practically the same in all indices. Analogous results were also obtained among the progeny of trispecific combination (dihaploid *S. andigenum* × *S. stenotomum*) × *S. pampasense*.

The above-mentioned hybrid forms were found to be highly fertile and from their self-pollination we got and studied more than 1500 lines of the I_1, I_2, I_3, out of which clones have been isolated with high yield potential and good tuber qualities. It can be seen from the data of Table 2 that the isolated lines maintain their high yield potential, high content of starch and protein and *Phytophthora* resistance in the I_1 and I_2. Some of the lines are of interest as carriers of a high percentage of protein (274/28–3; 274/28–15; 234/18–33; 239/42–1; 239/46–9). Apart from this the lines 239/14–6, 239/42–1, 239/46–9, 274/28–5 and 274/28–1–15 are early maturing lines; every year they produced tubers with a weight ranging from 30 to 60 g and a yield from 300 g (July 24) to 700 g (August 12) per plant. It must be mentioned that each line has many desirable traits.

The isolated lines were used by us in inter-line and mainly in varietal-line

Table 1. Yield potential and quality of the tubers in initial populations of F_2 and I_1

No. of initial population	Origin	Total number of clones	Mean ($\overline{X}$): Yield per plant, g		Mass of one tuber, g		Percentage			Coefficient of variation (V), %: Yield		Mass of one tuber		marketable tubers	protein	starch
			total	marketable	total	marketable	marketable tubers	protein in tubers	starch in tubers	total	marketable	total	marketable			
238/1	Dihaploid *S. andigenum* × *S. stenotomum*, F_2	7	963	869	62	88	90.8	1.8	17.8	6.4	10.9	35.7	13.6	2.4	4.6	11.7
238/1–3	Same, I_1	7	961	959	74	92	93.1	1.8	18.6	22.5	40.7	26.5	19.1	6.3	7.0	20.2
239/18	Dihaploid *S. andigenum* × *S. rybinii*, F_2	14	683	567	49	76	83.5	1.9	15.7	10.7	44.9	31.2	24.9	10.4	7.5	30.1
239/18–19	Same, I_1	6	290	210	45	64	—	—	16.6	52.9	20.2	10.3	0.6	—	—	14.1
274/28	238/7 × *S. pampasense*, F_2	15	664	578	55	76	84.1	2.2	16.9	33.7	35.5	27.6	14.3	9.9	12.7	9.9
274/28–3	Same, I_1	6	764	599	42	72	82.9	2.2	16.8	53.1	34.3	29.4	12.1	5.1	3.8	13.1

Table 2. Agro-economically valuable characters of the isolated lines

No. of inbred line	Progeny	Origin	Yield per plant, g		Mass of one tuber, g		Marketable tubers, %	Content in tubers, %		Phytophthora infection of haulms (estimated in points)	Taste qualities (estimated in points)
			total	marketable	total	marketable		protein	starch		
239/14–6	I_1	Dihaploid *S. andigenum* × *S. rybinii*	944	795	59	74	80.0	1.99	18.2	1	3.1
239/18–33	,,	,,	1021	919	69	84	87.5	2.19	14.8	0	2.2
239/42–1	,,	,,	1053	862	72	82	83.3	2.47	14.2	1	3.2
239/46–9	,,	,,	1122	768	65	90	86.7	2.13	15.5	2	2.6
239/18–09–10–1	I_3	,,	1052	873	81	114	83.0	1.85	17.2	1	
274/28–3	I_1	Dihaploid *S. andigenum* × *S. stenotomum* × *S. pampasense*	845	800	55	76	83.0	2.11	17.4	1	3.6
274/28–5	,,	,,	970	815	61	71	79.0	2.01	15.7	2	2.8
274/28–15	,,	,,	1141	967	70	92	84.8	2.05	16.7	0	3.2
274/28–1–15	I_2	,,	1031	842	74	106	78.8	1.78	11.5	1	
274/28–1–25	,,	,,	1076	710	55	81	85.7	1.18	17.6	0	
238/1–1–22	,,	Dihaploid *S. andigenum* × *S. stenotomum*	944	696	54	83	73.8	1.68	18.5	0	
238/1–3–6	,,	,,	861	715	56	82	81.3	1.88	16.6	1	

Table 3. A comparative evaluation of hybrid combinations from varietal-line and inter-line crosses

Combination	Total number of clones	Out of them in %							
		Yield per plant, kg		Marketable mass of one tuber of 100–200 g	Marketable tubers 90%	Protein in tubers 1.5%	Starch in tubers		With field resistance to Phytophthora
		1–1.5	More than 1.5				15–20%	20–25%	
Wilja × 239/18–9	5	60	20	80.0	80.0	40.0	0	0	59.0
Wilja × 239/18–33	10	50	20	60.0	70.4	0	20.0	0	20.0
Wilja × 239/18–33	7	57	0	57.2	42.4	0	57.2	0	42.9
Greta 8–101 × 239/18–33	6	33	16	50.0	66.4	0	33.2	0	83.3
Greta 8–92 × 239/18–33	8	20	0	25.0	40.0	50.0	0	0	87.5
Agora × 239/18–9	6	33	0	66.4	83.0	50.0	83.0	0	49.8
Primerosa × 239/18–33	9	77	0	55.5	77.7	11.0	55.5	0	55.5
Fanal × 239/18–33	4	10	25	50.0	25.0	0	100.0	0	75.0
Zoo 179 × 239/18–33	5	20	0	40.0	40.0	0	0	0	100.0
239/18–33 × Tunika	20	40	0	40.0	25.0	0	65.0	0	80.0
239/18–34 × Tunika	8	37	0	12.5	12.5	—	—	—	—
274/28–1–25 × Tunika	11	0	0	9.0	18.2	0	54.0	54.0	54.0
274/28–4 × Zette	2	50	0	0	50.0	—	—	—	50.0
239/14–6 × 274/28–10	7	28	0	28.6	14.0	28.6	57.2	57.2	100.0

Table 4. The characteristics of some specimens of varietal-line hybrids (1974–1976)

No. of the specimen	Hybrid	Yield per plant, g		Mass of one tuber, g		Marketable tubers, %	Content in tubers, %		Phytophthora infection (estimated in points)	
		total	marketable	total	marketable		protein	starch	haulms	tubers*
537/4	Wilja × 239/18–9	1230	1147	88	105	96.4	1.71	13.6	1	4
538/1	Wilja × 239/18–33	1517	1440	96	112	95.7	1.32	10.1	2	3
538/8	,,	976	909	96	116	95.9	1.29	11.3	1	2
539/1	Agora × 239/18–9	1299	1215	105	128	93.5	1.22	13.9	0	3
540/2	Primerosa × 239/18–33	1288	1190	108	121	97.8	1.68	16.8	2	2
541/5	Wilja × 239/18–33	1139	1098	117	128	97.7	1.00	13.9	0	3
542/3	Lara × 274/28–4	1155	912	48	76	84.8	1.37	15.6	0	2
543/2	Lara × 239/18–33	1015	912	75	95	88.1	1.59	12.4	0	26 (without spore formation)
581/3	Sunia × 274/28–15	1678	1536	73	92	91.3	1.34	14.3	0	4

*Under artificial infection.

hybridization with the development of promising varieties for the GDR. The Netherlands, Poland, Finland and Mexico. The combining ability of the individual lines was revealed as a result of hybridization, and better hybrid combinations were identified. The highest combining ability was attained by the line 239/18–33 which showed positive results with all the varieties used in the crosses. Good results were also obtained from crosses involving the line 239/18–9. Among all the combinations under study we identified 14 better ones, 8 of which involved the line 239/18–33 (Table 3).

From the combinations shown in Table 3, we selected individual hybrid specimens (clones) for further trials. Preliminarily, we can mention the following as comparatively better clones: No. 537/4, 540/2, 541/5, 543/2 and others (Table 4). Some of these require further trials regarding certain characters. At present work is being continued on the isolation of combinations as a whole, as well as on individual clones.

Results of the crossings of inbred lines with varieties, and also between inbred lines, confirm the effectiveness of this method for developing valuable initial material as a basis for future varieties. An advantage of such hybridization lies in the fact that we are able to isolate prospective forms from a small volume of population. For example, within five years we could get 85 hybrid combinations (inter-line and varietal-line) consisting of 4000 saplings. And even from this small quantity we could identify 10 which have certain economic importance. Further study of hybrid inbred lines in potato breeding will open up even greater possibilities for its use in interspecific hybridization.

In this way, development of inbred lines of interspecific hybrids of the potato is a promising way to get prospective forms for varietal-line and interline hybridization. Inbred lines can be valuable for one of many economic characters. With their use in practical breeding there is a greater probability of the isolation of agro-economically valuable clones from a comparatively small volume of hybrid material.

REFERENCES

1. Bukasov, S.M. and A.Ya. Kameraz. 1972. Selektsiya i Kartofelya (Breeding and seed technology of the potato). Leningrad.
2. Dospekhov, B.A. 1965. Metodika Polevogo Opyta (Method of Field Experimentation). Moscow.
3. Savchenko, A.P. and R.G. Kazak. 1975. Sozdanie intsukht-linii kartofelya s vysokim soderzhaniem krakhma la i ustoichivykh k fitoftore (The development of inbred lines of potatoes with high starch content and resistance to Phytophthora). In the book: *Issledovaniya po Teoreticheskoi i Prikladnoi Genetika*, Minsk.
4. Yashina, I.M. 1972. Programma i metodika izucheniya geneticheskoi prirody vazhneishikh khozyaistvennobiologicheskikh priznakov u kartofelya

i razrabotka effektivnykh metodov ispol'zovaniya etikh priznakov v selektsii (The program and method of study of the genetic nature of important agro-economic and biological characters of the potato and development of effective methods to use these characters in breeding). Moscow.

5. Krantz, F.A. 1929. Potato breeding methods. II. Selection on inbred lines. *Univ. Minnesota Agr. Exp. St. Techn. Bull.*, 58.
6. Krantz, F.A. 1946. Potato breeding methods. II. A suggested procedure for potato breeding. *Univ. Minnesota Techn. Bull.*, 173.
7. Krantz, F.A. 1951. Potato breeding in the United States. *Z. Pflanzenzuchtung*, 29, p. 3.

TRUDY PO PRIKLADNOI BOTANIKE, GENETIKE I SELEKTSII

1978 Vol. 62, No. 1

Eradication of Virus Diseases from Potato Plants in the VIR Collection by Meristem Tissue Culture: Some Results

E.V. Truskinov
Candidate of Agric. Sciences

From 1972 to 1976, eradication of virus diseases from potato specimens in the collection was carried out using the method of apical meristem culture. As a result, 120 specimens, representing different species, varieties, hybrids and clones of the potato, were made virus free.

The genetic heterogeneity of the material taken for eradication was responsible for: a) the variable response of the specimens to tissue culture, b) the maintenance of isolated meristems, c) the number of regenerated plants, d) the dates of their development into complete plants, e) the effect of eradication. Positive results were attained more easily with some species or varieties than with others.

The severe occurrence of virus diseases in the potato collection maintained at the VIR (All-Union Institute of Plant Industry) is causing serious damage to the world germplasm collection of this important crop. Such a situation has developed because of the fact that virus diseases are not listed as quarantine diseases and are imported with the mass of collection specimens. A study of imported tuber material in quarantine shows that over a number of years, on an average, more than 60% of the potato specimens imported from abroad were infected by a virus. In the last few years there has been a considerable increase in the percentage of specimens infected with yellow mycoplasmic disease, particularly among the expedition material brought from South America.

As an active factor possessing extremely infectious and pathogenic properties, the viruses are particularly virulent for the collection crops in the field, infecting more than 90% of the tuber specimens in a number of cases [5]. In this context it is urgently necessary to work out effective control measures, prophylactic as well as therapeutic, against virus diseases in the collection. Some principles of seed technological work on the potato against the vectors, etc., are fully acceptable, although not fully satisfactory, in sanitation eradication of virus infection from the collection. In many cases they do not give the desired results and then the only possible and the most radical sanitation eradication method is tissue culture and other therapeutic methods [2, 4, 9, 12].

In 1972, for the first time work was initiated at the VIR on making the potato free from virus by these methods. As a result, the variety Detskosel'skii is now free from viral infection, followed by the sanitation purification of many other valuable collection specimens [6, 7]. Since 1974 this work has assumed a systematic and massive nature. During the period 1972–1976, about 300 specimens representing different species and improved varieties were treated to eradicate the viruses. Regenerated plants have been obtained in about 50% of the specimens, of which 40% (about 120 specimens) have been made free from virus.

The specimens thus treated (see Table) reacted differently to quite homogeneous tissue culture conditions in our work: the same nutrient culture of Murashige-Skuga and the same place and method of culture, under a light apparatus with a 16-hour photoperiod, 4000–6000 luxes at an average temperature of 20–25°C was used throughout. Maximum fluctuations in the environment occurred in the temperature as well as in the seasonal changes which are significant natural factors in the differing reaction of plants to tissue culture [10]. Under the conditions of our experiments, regeneration was favorably affected by the spring-summer period which was obviously linked with the natural physiological growth processes of plants at that time.

However, the most important reason for the differing reaction of meristematic explants was obviously the great genotypical heterogeneity of specimens selected for purification representing different species, improved varieties, hybrids and clones of potato. To some extent, this could have affected the survival of the isolated apical meristems which varied from specimen to specimen from 12.5% to 100%, but to an even greater extent it affected the emergence of regeneratad plants (from 2.4 to 50%) and the time of their formation, from 30 days to one year or more.

Here it must be emphasized that only about half the specimens used in the work gave regenerated plants. Even after several efforts we could not get the

Results of the purification of potato collection specimens from virus diseases by the method of apical meristem culture (Pavlovskaya Experiment Station of the VIR, 1972–1976)

Catalogue number of VIR	Specimen	Number of isolated meristems	Infection of meristems		Plants obtained from the isolated meristems				Duration of regeneration, days*
					total		without viruses		
			absolute	%	absolute	%	absolute	%	
1	2	3	4	5	6	7	8	9	10
11928	*S. chacoense* Bitt.	11	5	45.5	1	9.1	1	9.1	157
11929	,,	7	2	28.6	1	14.3	1	14.3	60
12168	*S. immite* Dun.	8	4	50.0	2	25.0	2	25.0	262
7600	*S. sykinii* Lechn.	11	3	27.4	1	9.1	1	9.1	70
11288	,,	18	9	50.0	7	33.3	5	22.2	65
11289	*S. ochoanum* Lechn.	7	1	14.3	1	14.3	1	14.3	92
9384	*S. phureja* Juz. et Buk.	36	14	88.9	2	5.6	2	5.6	89
13707	,,	5	4	80.0	1	20.0	1	20.0	91
8211	*S. kesselbrenneri* Juz. et Buk.	37	5	13.5	1	2.7	1	2.7	101
9529	,,	30	18	60.0	3	10.0	3	10.0	138
11291	,,	10	9	90.0	2	20.0	2	20.0	150
10945	*S. goniocalyx* Juz. et Buk.	8	3	37.5	2	25.0	2	25.0	162
9875	*S. stenotomum* Juz. et Buk.	40	11	27.5	4	10.0	1	2.5	38
9889	,,	46	22	47.8	15	32.6	11	23.9	38
9906	,,	70	32	45.7	19	27.1	12	17.1	38
10479	,,	24	16	66.7	12	50.0	8	33.3	—
11090	,,	16	12	75.0	2	12.5	2	12.5	126
12779	,,	4	3	75.0	1	25.0	1	25.0	189

11003	*S. chocclo* Buk. et Lechn.	16	12	75.0	2	12.5	2	12.5	131
12684	,,	4	3	75.0	2	50.0	2	50.0	167
13451	,,	5	5	100.0	1	20.0	1	20.0	112
13303**	*S. chaucha* Juz. et Buk.	15	11	73.3	3	20.0	1	6.7	136
12922	*S. tenuifilamentum* Juz. et Buk.	7	6	85.7	2	28.6	2	28.6	215
12927	,,	6	4	66.7	1	16.7	1	16.7	249
13186	,,	5	4	80.0	1	20.0	1	20.0	117
13405	*S. juzepczukii* Buk.	4	3	75.0	1	25.0	1	25.0	107
11105	*S. curtilobum* Juz. et Buk.	12	5	41.7	1	8.3	1	8.3	248
13198	,,	8	4	50.0	1	12.5	1	12.5	201
10804	*S. andigenum* Juz. et Buk.	11	7	63.6	1	9.1	1	9.1	308
10940	,,	8	2	25.0	1	12.5	1	12.5	83
10957	,,	12	4	33.3	2	16.7	2	16.7	84
10963	,,	10	8	80.0	4	40.0	3	30.0	173
11037	,,	6	3	50.0	1	16.7	1	16.7	133
11736	,,	13	9	69.2	3	23.0	3	23.0	110
12727	,,	5	4	80.0	1	20.0	1	20.0	207
12813	,,	9	8	88.9	2	22.2	1	11.1	142
12828	,,	9	6	66.7	2	22.2	2	22.2	259
12910	,,	7	3	42.9	1	14.3	1	14.3	80
13005	,,	11	6	55.6	3	27.3	1	9.1	123
13198	,,	18	5	27.8	2	11.1	2	11.1	172
13567	,,	8	6	75.0	3	37.5	3	37.5	66
13645	,,	7	4	57.1	2	28.6	2	28.6	220
13800	,,	15	8	53.3	1	6.7	1	6.7	82
13807	,,	4	3	75.0	1	25.0	1	25.0	231
14077	,,	18	10	55.5	4	22.2	3	16.7	138
15420	,,	11	5	45.5	1	9.1	1	9.1	64
15443	,,	8	6	75.0	1	12.5	1	12.5	123
15476	,,	10	6	60.0	2	20.0	2	20.0	159
15755	,,	8	8	100.0	4	50.0	4	50.0	62
16053	,,	8	3	37.5	2	25.0	2	25.0	145

(*Contd.*)

1	2	3	4	5	6	7	8	9	10
16243	*S. andigenum* Juz. et Buk.	8	6	75.0	2	25.0	2	25.0	66
16261	,,	9	5	55.5	3	33.3	3	33.3	124
3400***	*S. chilotanum* Hawk.	15	10	66.7	2	13.3	2	13.3	230
3403	,,	26	13	50.0	3	11.5	2	7.7	179
3407	,,	28	13	46.4	4	14.2	2	7.1	135
3455**	,,	11	5	45.4	1	9.1	1	9.1	342
3484	,,	28	13	46.4	6	21.4	6	21.4	166
3499	,,	59	32	54.2	7	11.9	3	5.1	260
3500	,,	49	8	16.3	5	10.2	5	10.2	84
3501	,,	22	9	40.9	1	4.5	1	4.5	196
3502	,,	40	24	60.0	7	17.5	7	17.5	60
6004***	,,	14	—	—	3	21.4	3	21.4	244
6108	,,	100	26	26.0	4	4.0	1	1.0	61
7530	,,	30	21	70.0	5	16.7	2	6.7	123
7596	,,	24	16	66.7	1	4.3	1	4.3	244
10601	,,	24	8	33.3	4	16.7	4	16.7	77
10613	,,	18	5	27.8	1	5.6	1	5.6	89
10635	,,	26	7	26.9	3	11.5	2	7.7	103
10637	,,	30	17	56.7	4	13.3	3	10.0	60
10638	,,	20	12	60.0	1	5.0	1	5.0	181
10646	,,	30	6	20.0	3	10.0	2	6.7	145
10647	,,	41	10	24.4	1	2.4	1	2.4	128
10654	,,	39	21	53.8	1	2.6	1	2.6	271
B_4–825	,,	22	14	63.6	3	13.6	2	9.1	52
B_4–911, No. 386	,,	16	6	37.5	1	6.3	1	6.3	203
B_4–9, No. 1916***	,,	23	11	47.8	2	8.7	1	4.3	263
	Improved varieties of *S. tuberosum* L.								
2902	Detskosel'skii**	500	—	—	20	4.0	19	3.8	75
6886	Veselovskii 2–4	50	34	68.0	7	14.0	3	6.0	102
10127	Iskra	23	14	60.9	3	13.0	1	4.3	246

10128	Istrinskii	21	16	76.2	2	9.5	2	9.5	168
6914	Povirovets	28	20	71.4	3	10.7	2	7.1	75
11510	Avanti	18	10	55.6	3	16.7	2	11.2	84
15006	Bintje	10	9	90.0	3	30.0	3	30.0	153
14989	Birga	16	11	68.8	1	6.3	1	6.3	263
11559	Conny	14	9	64.2	3	21.4	1	7.1	99
15028	Gelda	11	9	81.8	3	27.3	2	18.2	105
11614	Granit	15	7	46.7	1	6.7	1	6.7	154
14993	Hansa	4	2	50.0	1	25.0	1	25.0	295
5025	Heiko	8	1	12.5	1	12.5	1	12.5	168
12857**	Hudson	11	6	54.5	1	9.1	1	9.1	240
14998	Isabell	15	14	93.3	1	6.7	1	6.7	219
14988	Kamyk	11	3	27.3	1	9.1	1	9.1	40
12020	Kufri Lauvkar	16	8	50.0	3	18.7	2	12.5	87
11505	Kufri Neelmani	19	14	73.7	1	5.3	1	5.3	215
14999	Linda	8	5	62.5	1	12.5	1	12.5	205
11572***	Lindenhof	19	18	94.7	3	15.8	3	15.8	330
12858	Nampa	12	5	41.7	4	33.3	2	16.7	145
11615	Notec	9	5	55.6	1	11.1	1	11.1	127
12113	Nyirsegi rozsa	21	10	47.6	5	23.9	4	19.0	77
11556	R. Sasumna	10	9	90.0	1	10.0	1	10.0	126
14980	Stania	6	6	100.0	2	33.3	2	33.3	187
14981	Vokal	10	8	80.0	1	10.0	1	10.0	367
12861	Sapling USDA 96–56	9	3	33.3	2	22.2	1	11.1	92
12862	,, ,, 41956	5	2	40.0	1	20.0	1	20.0	318
11610	Dihaploid Ag. M 5971	13	10	76.9	3	23.1	3	23.1	66
	Spartaan								
15008	,, VI 11028 Mariella	8	4	50.0	2	25.0	2	25.0	161
12867	Hybrid M–12	8	5	62.5	1	12.5	1	12.5	102
12868	,, M–16	7	6	85.7	3	42.8	1	14.3	102

*Time from isolation of meristem to planting of the plant in the soil.
**Thermotherapy used before meristem culture.
***Chemotherapy used before meristem culture.

desired plants from specimens of some species such as *S. ajanhuiri* Juz. et Buk., while the specimens of *S. stenotomum* Juz. et Buk. almost always responded well to apical meristem culture. Generally the specimens of *S. andigenum* Juz. et Buk. regenerated well on the nutrient culture (9.1–50% of the plants), whereas specimens of *S. chilotanum* Juz. et Buk. gave a poor response (2.4–21.4%).

The link between genotype and the effect of purification was less evident. However, the results obtained by us show indirectly that such a link certainly does exist. It is not possible to liberate some specimens from the viruses even using an isolate of a size up to 0.1 mm under recurrent cycles of eradication, while others can comparatively easily be made free from virus infection using meristems measuring 0.25–0.30 mm. This refers even to cases of infection by virus S which is very difficult to eradicate.

In some cases, when tissue therapy did not give the desired results, thermotherapy was resorted to as a supplementary method. If thermotherapy of the plants themselves was used in the variety Detskosel'skii, then in the case of other specimens it was done with the tubers and selected plants. While doing so we were confronted with difficulties in the fairly limited, yet not uniform, endurance of the potato plants and tubers to high temperatures. Nevertheless, a combination of thermotherapy with tissue culture proved highly effective for some specimens. Apart from this, we conducted several experiments on chemotherapy using different virus inhibitors and different methods of application. Although at present it is not possible to speak of any reliable results, we can assume that the addition of certain inhibitors to the nutrient culture did play a definite role in the purification of some specimens.

The use of certain chemicals in the inactivation of viruses under conditions of tissue culture is of great interest. This is because, according to some data, viruses are present in the meristem zone anyway [1, 11]. Therefore, it is quite logical to assume that in tissue culture we always have a chemotherapeutic side-effect, especially when the majority of growth substances used in the nutrient culture are simultaneously inhibitors of viruses. In this case, the whole problem is only to intensify this effect by the addition of the corresponding chemical without causing any damage to the regeneration of normal plants.

Tissue culture itself, as well as in combination with thermo- and chemotherapy, can be the reason for the functional and structural variability of regenerated plants in a number of cases. We have already recorded certain anomalies in the development of explants under conditions of tissue culture [8]. While a study of such regenerated plants and their botanic-taxonomic identification has not been conducted yet, we have certain data on the existence of variability in individual specimens. So far, without knowing anything about the nature of such variability, it must be said that in many cases the modified variability (not hereditary) could be caused by the purification treatment. This may relate to the growth habit as well as some physiological properties of the plants. Moreover, the possibility of mutation (particularly in the formation of callus) cannot be excluded, although mutation may not occur with a higher frequency than under natural conditions. If variation is caused by purification treatments, then the entire work on purification by the method of meristem tissue culture would have become impossible and would only have importance in breeding.

Eradication of viruses from plants through tissue culture, as we know, does not give them any resistance to further virus or any other infection. On the contrary, there is data on the greater susceptibility of virus-free material to different pathogens including viruses. It is thoroughly understandable in light of the phenomenon of cross-protection, which obviously takes place not only at the intraspecific level of the strain but in a number of cases at the interspecific level of the pathogens [3].

On the basis of our experiments it can be confirmed that virus-free material obtained from the meristem was reinfected quickly (in the field as well as in greenhouses) when there were no effective control measures against viral infection. Vectoral transfer of infection by winged aphids and other insects poses a particular danger. According to our observations, under such circumstances the leaf curling virus is most widely dispersed. However, in some cases it was not always possible to confirm the infectious (viral) origin of the symptoms of leaf curling, even with the help of indicator analysis (infection of test-plants by the infected aphid *Physalis floridana*). At the same time it is essential to mention that, even without any reliable protection of the meristem plants against reinfection, there are specimens which have remained virus-free for a period of three to four years. Use of more effective means of isolation of the virus-free material (separate greenhouses, isolation cabins, as well as natural secluded zones) and regular control of the viral vectors are the fundamental measures ensuring prolonged maintenance of the collection material in a virus-free condition. Along with this it is very important to maintain the mother collection of specimens which have been made virus-free through the method of tissue culture. This collection should be maintained in extremely reliable isolation so that it can serve as a source of constant rejuvenation of the virus-free collection specimens in the field and in the greenhouses. We have initiated such a collection for a number of specimens and have maintained it for many years.

From the above account it follows that a purification treatment of potato collection specimens by the method of meristem tissue definitely gives positive results, although they cannot be guaranteed. It requires great care to maintain the purified material in such a condition for a long time. At the same time, despite its universal application and effectiveness, tissue culture as a method of eradicating the viruses in plant material cannot and must not replace the remaining means of eradication (for example, wherever possible, clonal selection, reproduction through seeds, etc.) and it should find an appropriate place in the system of purification measures. This method should be appropriately used for specimens which have 100% virus infection and which die without producing seeds. First of all, this pertains to species and varieties as well as valuable hybrids and clones which do not produce seeds. Considering the huge volume and diversity of the collection as well as its high degree of infection by all possible types of viruses, strict selection of material for purification by the method of tissue culture is absolutely essential.

CONCLUSIONS

1. As a result of the work on eradicating virus diseases from the potato collection specimens by the method of apical meristem tissue culture, conducted from 1972 to 1976 at the Pavlovskaya Experiment Station of the VIR, about 120 specimens of different species, improved varieties, hybrids and clones of the potato have been made free from viruses.

2. In the course of this work the different reactions of genetically diverse specimens of potatoes to the conditions of meristem tissue culture were revealed. Quantitatively this was expressed in the different degrees of survival of the explants and in the regeneration of plants as well of their degree of purification. Some qualitative differences at the species level were also observed, which allowed us to get regenerated plants comparatively easily from species such as *S. stenotomum* Juz. et Buk. and *S. andigenum* Juz. et Buk., not so easily from *S. chilotanum* and without any positive result from *S. ajanhuiri* Juz. et Buk.

3. Material made virus-free by the method of tissue culture is liable to a fairly high degree of reinfection and thus it requires a special system for protection against reinfection. The most reliable method for maintaining virus-free material is its permanent maintenance on a nutrient culture in test tubes, ensuring isolation of plants from infection as well as rejuvenation of specimens in the field (in case of their infection or death).

4. Considering the huge volume and extreme diversity of the potato collection, and also the high degree of its virus infection, purification of the collection by the method of tissue culture should be done first on samples which have 100% virus infection in a highly virulent form and which cannot produce fully viable seeds for any purpose (some species, all the man-made improved varieties, hybrids and clones.

REFERENCES

1. Krylova, N.V. and V.I. Stepanenko. 1971. Pronikayut li virusy v apikal'nuyu meristemu (Do viruses penetrate the apical meristem?). In the book: *Virusnye Boleznii S. Kh. Rastenii Dal'nego Vostoka*, Vol. 3, p. 255.
2. Morel', Zh. 1967. Bor'ba s virusnymi boleznyami (rastenii s pomoshch'yu

kul'tivirovaniya tkanei (Control of virus diseases of plants by tissue culture). *Sel'khoz. Biologiya*, Vol. 2, No. 4, pp. 622–628.

3. Met'yuz, R. 1973. Virusy Rastenii (Viruses of plants). Izd-vo "Mir", Moscow, 600 pages.
4. Trofimets, L.N., V.A. Knuyazev, L.M. Khromova and D.P. Ostapenko. 1976. Metodicheskie Ukazaniya po Ozdorovleniyu i Uskorennomu Razmnozheniyu Kartofelya (Methodical instructions on purification and quick multiplication of the potato). VASKhNIL, NIIKKh, Moscow.
5. Truskinov, E.V. 1972. Porazhenie mirovoi kollektsii Kartofelya mozaichnymi virusami i perspektivy selektsii na ustoichivost' k virusu (Infection of the world potato collection by mosaic viruses and prospects of breeding for virus resistance). Abstract of a Ph.D. thesis. Moscow.
6. Truskinov, E.V. 1975. Osvobozhdenie kartofelya sorta Detskosel'skii ot virusa S (Purification of the potato variety Detskosel'skii from virus S). *Kartofel' i Ovoshchi*, No. 10, pp. 33–34.
7. Truskinov, E.V. 1976. Kul'tura tkanei kak sredstvo bor'by s virusnymi boleznyami introdutsirovannogo kartofelya (Tissue culture as a control measure against virus diseases of the introduced potato). *Byull. VIR*, No. 61, pp. 58–61.
8. Truskinov, E.V. 1976. Ozdorovlenie kartofelya ot virusnykh boleznei metodom kul'tury meristemnykh tkanei (Purification of the potato from virus diseases by the method of meristem tissue culture). *Sel'khoz. Biologiya*, Vol. 11, No. 2, pp. 250–255.
9. McDonald, D.M. 1973. Heat treatment and meristem culture as a means of freeing potato varieties from viruses X and S. *Potato Research*, Vol. 16, No. 4, pp. 263–269.
10. Mellor, F.C. and R. Stace-Smith. 1969. Development of excised potato buds in nutrient culture. *Canad. Journal of Botany*, Vol. 47, No. 10, pp. 1617–1621.
11. Pennazio, S. and P. Redolfi. 1974. Potato virus X eradication in cultured potato meristem tips. *Pot. Res.*, Vol. 17, No. 3, pp. 333–335.
12. Quak, F. 1961. Heat treatment and substances inhibiting virus multiplication, in meristem culture to obtain virus-free plant. *Adv. Hort. Sci. Applic.*, Vol. 1, pp. 144–148.

TRUDY PO PRIKLADNOI BOTANIKE, GENETIKE I SELEKTSII

1978 Vol. 62, No. 1

Measures to Increase the Efficiency of Primary Seed Production of Virus-free Potatoes

A.G. Zykin
Doctor of Agric. Sciences

In the primary seed production of potatoes, the use of virus-free initial material purified by laboratory methods is better in all respects. It requires a significantly longer time, and greater labor, material and means to obtain such virus-free material from plantings or clone nurseries, with a high degree of purification of viruses. In some varieties these attempts have failed altogether.

Multiplication of virus-free initial material under a gauze cover helps maintain such material in a healthy state for a long time. If one tuber from each plant grown under these conditions is used for planting a new protected crop the following year and the other tubers are planted in the field, production of healthy clones is ensured every year.

Cutting the tubers into two or more parts before planting increases the coefficient of clone multiplication 1.5–2 times. This greatly decreases both the demand for initial virus-free material and the expenses for growing them.

The coefficient of multiplication of the clones depends on the plant density of the tubers cut into parts: the maximum coefficient is reached in sparse plantings by the two-strip method, whereas the minimum is in close-drill plantings.

Planting the clones by the 2-strip method makes it possible to use the common 2-drill potato diggers for harvesting them.

The potato is infected by a few dozen virulent diseases which is why there is a considerable loss of yield and deterioration in the market quality of the tubers. In the main potato-growing regions of our country the most common and virulent are Phytophthorosis (late blight) and virus diseases.

The majority of the pathogens are transmitted through planting material from one vegetative generation to another, accumulating it in ever increasing amounts. Therefore, an organization to grow healthy seed potatoes and their use in commercial plantings are the most reliable guarantees for a high and stable yield of this valuable agricultural crop.

The cultivation of healthy seed potatoes is linked with a series of difficult and costly measures. It requires complicated organization to protect the potato against virus diseases, since the spread of their pathogens is distinguished by exceptional diversity. Individual viruses and their strains can be spread by different terrestrial vectors or carriers (aphids, leafhoppers, field bugs and biting insects), some species of parasitic soil fungi and nematodes, as well as through contact with mechanically injured organs and tissues of the plants and tubers. Moreover it is spread from the tuber seed of one vegetative generation to the next [2].

Under field and storage conditions it is practically impossible, and economically very costly, to check the introduction of virus infection into the potato and its spread inside the potato. As a result of this, during the course of the breeding process or in the initial stages of the multiplication of potato varieties, there invariably occur isolated diseased plants and tubers whose numbers continuously increase with every new tuber production. Some potato varieties are infected by one or more viruses in the sapling nursery, i.e. in the first year of its existence.

In the first case initially the high yield potential of varieties becomes drastically reduced and in the second case their yield potential does not manifest itself to the full extent. This is the reason why many valuable potato varieties were either not released for cultivation or were used for commercial cultivation for only a very brief period. Large-scale commercial crops of the released varieties were 100% infected by a complex of latent viruses, while a considerable number of plants had severe symptoms of various diseases. In this context, according to our calculations, the annual average loss in the yield of potato in our country is 15–20 million tons, valued at 1.5–2 billion rubles according to the retail price of potato [1, 2].

Thus there is an urgent need to organize the cultivation of virus-free seed material. This is an indispensable condition for prolonging the life and maintaining the high yield potential of potato varieties and it is also the main measure for increasing yield potential.

So far scientists have worked out and recommended to the potato growers a whole complex of measures to protect the seed potato from virus disease infection. This complex includes such measures as space (distance), isolation of seed nurseries from the sources of virus infection and its vectors, pre-planting germination of the seed tubers, dense planting, treatment of tubers and plants with systematic insecticides, several phyto-thinnings, thinning and other interculture operations till the haulms are fully grown, early harvesting with advance destruction of the haulms, grading of the freshly harvested tubers after their healing period and transportation and storage of the seed material in containers [2–4]. Some methods for accelerating the growth, development and tuber formation of the seed potato allow it to "skid" the periods when there are massive swarms of viral vectors, while others are directed towards the physical destruction or reduction in number and infection activity of the latter. Some measures ensure space (distance), and isolation from sources of virus infection or their timely removal from the potato-growing areas, whereas other measures prevent mass contact, and dispersal of disease pathogens through mechanical injuries in the plants or tubers and so on.

However, the above-mentioned measures, particularly when applied in a complex, are effective only in the control of primary infection, i.e. during the multiplication of seed potato with a low degree of infection of the disease pathogens. When growing seed material already highly infected with viruses in a latent form or when planting crops totally infected throughout their development, these measures lose their efficacy and they represent nothing but useless loss of labor and materials. Hence, the final result of many years' seed production work will depend foremost on the quality of the seed material grown at the beginning of the multiplication of potato varieties, i.e. on the primary seed production.

At present in our country as well as abroad primary seed production of the potato is done by the clonal method. This method is practiced in the following way. Every year from the most healthy crops of a particular variety plants are selected which are not infected by any virus infection, either in open or latent form, or by pathogens of other diseases. Clusters of tubers (clones) selected from these plants are additionally evaluated in winter under greenhouse conditions by the method of eye indexation, and then they are multiplied and tested over a period of two years or more, by growing each clone in the field in separate plots. Clones in which not a single diseased plant was found during evaluation are pooled together to form a batch while harvesting the nursery in the final year of the trial. Tubers of this batch are used to grow super-elite and elite seed stocks [1–4].

Until recently the only sources of initial material from which healthy progenitors of clones could be selected, were the most healthy crops of potato varieties existing on farms or in the research institutions where these varieties were developed. But as we have mentioned above, these are generally highly infected by viruses in an open as well as a latent form. Among such crops healthy plants are selected only in the first year; in subsequent years selection is done in the clonal nursery of the first year.

In the last few years more and more plant breeders have begun to use another source of initial material, i.e. tubers from plants grown under sterile conditions from the apical meristem. It is well known that initial material obtained by this method is free from virus infection as well as from pathogens of other potato diseases [1, 3].

In 1966–1977, for the primary seed production of the potato variety Detskosel'skii, under conditions at the Pushkin Laboratories of the VIR, we used nonpurified as well as purified initial material for the selection of clones (Table 1).

Table 1. Virus infection of the clones of first year potato depending on the quality of the initial material and the conditions for growth (1966–1977)

Year	Percentage of virus infection of clones while evaluating			
	by method of eye indexation		in the field	
	open	latent*	open	latent*
Open field	Selection from nonpurified initial material			
1966	—	—	43.5	85.0
1967	15.4	79.2	19.0	60.1
1968	3.9	19.7	15.1	7.9
1969	1.2	3.6	5.3	3.0
1970	6.4	4.5	2.7	1.8
1971	1.8	3.6	3.1	3.0
1972	2.3	1.1	6.5	2.1
1973	2.6	2.0	3.7	0.3
Open field	Selection from initial material purified by the method of meristem tissue culture			
1974	0	0	0	0
1975	0	0	0.6	19.8
1976	17.5	99.3	0.7	2.2
1977	5.1	7.5	13.8	10.7
Isolated under gauze cover				
1974	0	0	0	0
1975	0	0	0	0
1976	0	0.3	0.5	0.8
1977	0	0	0.6	0.9

*In 1966–1973 infection was estimated for all viruses except virus S, which totally infected the clones.

We could not select a single absolutely virus-free clone from the nonpurified initial material in which 43.5% of the plants were visually diseased (mostly by wrinkled mosaic and leaf curling virus), 100% were infected by virus S and

85% by other viruses in latent form. Several thousand plants were examined by us and all were infected by virus S. Comparatively healthy clonal material, infected by only one virus, could be selected only from the fourth year of selection. In the first two years more than half the selected clones were rejected during subsequent trials because of the presence of virus S and infections of other viruses in open or latent form in them. In the third year, according to evaluation by the method of eye indexation and in summer under field conditions, such diseased clones comprised about 50%, and in the fourth and subsequent years only 10–15%, of the total number.

In our opinion the above-mentioned facts can be explained by the following conditions. The methods used in the evaluation of clones for diagnosis of virus diseases and their pathogens are not totally reliable. Errors are always possible while conducting the evaluation and diagnosis of viruses. Therefore, some virus-infected clones invariably are selected along with the healthy clones. These clones infect the healthy ones during storage and field trials. The amount of permissible error and the intensity of the process of infection of healthy clones from the diseased ones are in direct correlation with the degree of virus infection of the initial material.

Hence, primary seed production of the potato, carried out by clone selection from preliminarily nonpurified initial material, is a prolonged, time-consuming and difficult process requiring a large volume of tedious phytopathological work over a period of many years. Furthermore, it does not provide an opportunity to select clones of certain varieties which are free from all known viruses.

In primary seed production switching over to the use of initial material purified from viruses and other pathogens of diseases by the method of apical meristem tissue culture enabled us to drastically raise the quality of the selected clones and also the elite seed stock of the potato which was raised from them. However, there emerged the problem of multiplication of virus-free initial material. It was found that it is not very difficult or costly under laboratory conditions to raise healthy plants from the meristem and to obtain some virus-free tubers from them. But it is highly complicated and difficult to raise the latter in the volume of comparatively large seed batches and to prevent them from massive new virus infections at the same time.

At the initial stages of multiplication of the initial virus-free material purified by the method of meristem tissue culture it is commonly grown under isolation conditions: in greenhouses, in beds under gauze or nylon covers and so on. But using such methods it is possible to raise only a very small quantity of planting material. Later on, we are compelled to multiply the planting material in an unprotected open field in small plots. It is well known that the smaller the size of the area of seed potato, the more intensive is the new virus infection of the plants [1–3]. Besides, the speed of this process depends on the susceptibility of the potato varieties to individual viruses, their remoteness from the sources of virus infection, the number and growing conditions of the plants and so on. Hence, organization of the multiplication of virus-free initial material must be planned with due consideration to the above-mentioned factors. In 1973–1977 we studied this problem at the Pushkin Laboratories of the VIR while multiplying virus-free initial material of variety Detskosel'skii, which had been purified by the method of meristem tissue culture at the Pavlovskaya quarantine greenhouse.

In the first year of multiplication the quantity of initial material was very small—only 21 tubers. These tubers were planted in the spring in the open field and the plants developing from them were grown under a gauze cover from the moment the sprouts appeared until harvesting. According to the data from several tests, the plants as well as the tubers collected from them were healthy.

Beginning in 1974, the virus-free material was divided into two parts; we continued to multiply one part under a gauze cover in special isolation beds and plants were sent to the first year clonal nursery for multiplication and trial (Table 1). The data presented in the Table shows that growing virus-free material under conditions in which they are unprotected against the vectors leads to loss of its valuable properties. New virus infection of the initial material was particularly noticed in 1975, when there was an unusually high number of aphids, the vectors of viruses. Out of 1500 clones grown that year under unprotected conditions only two were free of virus infection when they were examined in the winter by the method of indexation.

In the second version of multiplication and maintenance of initial material in gauze houses, during the first two years, including 1975, there was no virus infection but beginning with the third year isolated diseased plants began to appear. The extent of virus infection in clones, which were selected from the initial material grown under gauze cover, did not exceed 0.5–0.6 according to external symptoms and 0.8–0.9% in a latent form (mostly virus S).

In this way, multiplication of virus-free initial material under isolation conditions (under gauze or nylon nets) is comparatively more reliable and protects it against massive infection by the pathogens of diseases and enables us to get clones with a very insignificant degree of virus infection. Considering this, during the last three years, instead of selection of clones in the first year clonal nursery, we began to grow them in isolated gauze houses (see Figure). In order to have a continuous production process of clonal material, we used one tuber from every clone for planting the following year, again in the isolation beds. This is the reason why the portion of field occupied by isolation beds has been termed the "production nursery of virus-free initial clones".

Besides this, in order to avoid mass virus infection of the first year clones, we started their production and trial on the isolation plot with a special three-field crop rotation (annual grasses for green manuring, cereals and potato). The plot was 400–500 meters from other potato crops and was surrounded by a metallic net. All the working tools and implements, tractors and machines were washed with water and disinfected with formalin before starting work or before they entered the plot.

The first year clonal nursery of production by the double line-strip method of planting.

However, in clone cultivation in the isolation plots all the work relating to planting, upkeep of plants and harvesting are done manually. However, preparation and installation of the apparatus also require a comparatively large labor and material expenditure. And if we take into consideration the fact that four plants can be raised in 1 m^2 area of isolation plot, then for every 100 tons of elite tubers generally we have selected 800–1000 clones and in order to grow that quantity of elite stock every year we require 200–250 m^2 of area in the greenhouse. All this increases the cost of production of clonal material considerably and increases the labor requirements.

The search for ways to reduce expenditure in growing the initial material and in producing healthy clones resulted in the conclusion that this is possible by increasing the coefficient of multiplication of the clonal material. The higher the latter, the less would be the requirement of each clone and also the labor and material expenditure.

It is known that the most simple, economic and effective measure to increase the coefficient of multiplication of the seed potato is the cutting of planting tubers (with careful disinfection of the instruments being used for this purpose) into halves or smaller parts, each of them having not less than two eyes. For example, this measure is widely applied in Canada in primary seed-production. In some varieties this measure yields up to 1 ton of seed material from one initial virus-free tuber within a period of three years [5].

Use of pre-planting cuttings of seed tubers in the primary seed-production of the potato variety Detskosel'skii, under the conditions in Leningrad province, showed that an increase in the coefficient of multiplication of clonal material to a great extent depends on the density of the planting of the halves or parts of tubers, i.e. on the area of nutrition of the plants developing from them (Tables 2 and 3).

Table 2. The coefficient of multiplication of clonal material of the potato in planting by the row method (1975–1977)

Planting material*	Density of planting, cm	Yield of seed tubers, in number per plant	Coefficient of multiplication
1975			
Halves	70 × 30	8.6	17
Whole	70 × 30	13.5	13
1976			
Halves	70 × 30	7.2	14
,,	70 × 15	6.1	12
Whole	70 × 30	12.2	12
1977			
Halves	70 × 30	7.5	15
Whole	70 × 30	12.1	12

*Tubers of clones of the first and second year of multiplication with mass of 60–150 g, well germinated in the light.

Dense planting with halves of the tubers in the usual row method provides a 70 × 15 cm area of nutrition. This hardly affects the coefficient of multiplication of the seed material in comparison with the planting of the same quantity of whole tubers with a distance of 30 cm between them in the row. In this case the effect of cutting was favorably seen only as an increase in the output of uniform tubers of average size from a unit area.

Planting with the halves further apart (with a distance of 30 cm between them in the row) increased the number of tubers formed per plant to 7.2–8.6,

Table 3. Coefficient of multiplication of the potato with the cutting of different sized seed tubers into parts, and separation of their plantings by the line-strip method*

Average mass of planting tuber, g	Number of parts from 1 cut tuber	Density of planting in the line, cm	Collection of seed tubers, in number per plant	Coefficient of multiplication
1976				
450	7	30	6.8	47
450	without cutting	90	18.6	19
300	5	30	8.9	44
300	without cutting	90	17.4	17
150	3	30	8.4	25
150	without cutting	90	15.7	16
75–150**	2	30	11.2	22
75–150	without cutting	90	12.1	12
1977				
300	5	30	8.8	44
300	without cutting	90	25.8	26
150	3	30	9.7	29
150	without cutting	90	18.1	18
75–150**	2	30	10.5	21
75–150	without cutting	90	12.8	13

*Distance between strips 140 cm, between lines in the strip—70 cm.
**Average data for first year clonal nursery.

thus increasing the coefficient of multiplication to 14–17, i.e. 1.2–1.3 times that of planting with whole tubers. The yield per unit area was somewhat lower but this loss was compensated for to a great extent by the savings in the planting material.

The best results were obtained by cutting the average and big tubers into halves and the very big tubers into 3–7 parts, and then planting them by the double line-strip method with a distance of 140 cm between the strips, 70 cm between lines in the strip and 30 cm between the parts of tubers in the line (see Figure). With this method of planting, the plants form far more tubers because of the improvement in the light and nutrition conditions. This increases the coefficient of multiplication of the seed material, with the cutting of the planting tubers into halves, almost 1.5–1.7 times, and into several parts by two or more times. Accordingly there is a reduction of 4–5 times in the requirement of the initial clones and in the expenditure on their rearing in the isolation plots.

In 1976–1977 a wider application of the cutting of tubers into halves and the strip method of their planting led to an increase in the average number of plants in the first year clone of up to 35–40 and an increase in its yield of up to 300–320 tubers (against 12–13 plants and 140–150 tubers in the case of planting by the row method with dense planting material). There was a consider-

Table 4. Peculiarities in the growing of elite seed stock of potatoes o m virus-free initial material at the Pavlovskaya Experiment Station of the VIR (1975–78)

Year	Name of the nursery	Coefficient of multiplication	For each 100 tons of elite seed stock			Phytopurification and seed technological measures
			nutrition area, ha	quantity of tubers, in thousands		
				for planting	in the yield	
1	2	3	4	5	6	7
First	Production of virus-free initial clones	16	0.01	0.18	2.8	Planting with big tubers, treated with insecticides, germinated in moist turf and cut in half. Growing of plants in isolation plots, their treatment with insecticides and fungicides. Harvesting late. Examination of 25% of the tubers of each clone in winter by the method of eye indexation and use of one of them for planting the following year in the nursery.
Second	Trial of clones	18	0.3	2.5	45.0	Planting with tubers, germinated in the light, treated with insecticides and cut into halves or four parts, by the double line-strip method on an isolation plot with special three-field crop rotation. Rejection of clones which have diseased plants. Early harvesting with advance removal of haulms. Examination in winter by the method of eye indexation of two-three tubers of each clone.

(*Contd.*)

1	2	3	4	5	6	7
Third	Multiplication of clones	12	1.7	40.0	480.0	Germination and cutting of tubers into halves before planting. Planting in rows, dense in field crop rotation. Several phyto-rogueings. Harvesting early with advance removal of haulms.
Fourth	Elite seed stock	5	6.2	430.0	2150.0	Dense planting with whole germinated tubers. Several phyto-rogueings. Early harvesting with advance destruction of haulms.

able improvement in the conditions for conducting a visual evaluation to select leaf specimens for serological analysis. Furthermore the possibility of mechanical injuries to the vegetative mass of plants by man and machines could be excluded. Thus, this successfully solved the problem of mechanization for planting and harvesting the clones. For this purpose, half the seed tubers of every clone are placed in one line, and the other half in the second line of the strip at the time of planting.

Harvesting of clones planted by the above-mentioned method can be done successfully with the common two-row tractor-harvesters for potatoes. In order to avoid mixing tubers from succeeding clones with the tubers of previous clones, the potato harvester halts for a few seconds in between two clones during which time its elevators are freed from the mass of soil and potato.

In this way, the use of virus-free initial material, the production of progenitor clones in soil protected from viral vectors, growing clones of the first year of multiplication on isolated plots with special crop-rotation, increasing the coefficient of multiplication of clones by cutting the planting tubers into halves or smaller parts, planting first year clones by the double-line strip method, as well as the mechanized planting and harvesting of clones all considerably increase the efficiency of primary seed-production of potatoes. It is evident from the contents of Table 4 that the application of these measures at the Pavlovskaya Experiment Station of the VIR, along with other seed technological and phytopurification measures (additional examination of the clones in winter by the method of eye indexation, germination (sprouting) in the light and treatment of planting material with insecticides, timely detection and removal of invariably emerging diseased plants and clones from the nursery, early harvesting with advance destruction of haulms, etc.) all enable us to grow an elite seed stock of the variety Detskosel'skii on a virus-free basis in an abridged (4-year) scheme, using only 180 virus-free initial tubers for the production of each 100 tons of elite seed stock. For the past few years the station has been producing about 500 tons of elite seed stock of potato every year, and the degree of its infection by virus diseases does not exceed 1–2%.

REFERENCES

1. Zykin, A.G. 1974. Nauchnye osnovy vryashchivaniya semennogo kartofelya, svobodnogo ot virusov i mikoplazm (Scientific fundamentals of growing seed potato free from viruses and mycoplasm). Abstract of the D.Sc. dissertation. VIR, Leningrad, 54 pages.
2. Zykin, A.G. 1976. Virusnye Bolezni Kartofelya (Virus diseases of potato). Izd-vo "Kolos", Leningrad, 151 pages.
3. Adamov, I.I., L.N. Trofimets, A.G. Zykin, A.S. Volovik and E.I. Larina. 1974. Metodicheskie Ukazaniya po Semenovodstvu Kartofelya na Bezvirusnoi Osnove (Methodical instruction on seed production of the potato on

a virus-free basis). Moscow, 35 pages.

4. Polozhenie of Pervichnykh Pitomnikakh Vyrashchivaniya Elity Kartofelya (Present state of primary nurseries for growing elite seed stock of potatoes). 1971. "Kolos", Moscow, 23 pages.
5. Shmyglya, V.A. and A.G. Zykin. 1976. Semenovodstvo kartofelya v Kanade (Seed production of potatoes in Canada). *Kartofel' i Ovoshchi*, No. 6, pp. 45–46.

TRUDY PO PRIKLADNOI BOTANIKE, GENETIKE I SELEKTSII

1978 Vol. 62, No. 1

BRIEF REPORTS

Cultivated Tetraploid Forms of the Potato Species *Solanum andigenum* Juz. et Buk. as Initial Breeding Material for Nematode Resistance

E.V. Morozova
Candidate of Agric. Sciences

During the past few years the South American cultivated species *Solanum andigenum* has been used extensively for breeding varieties resistant to potato nematode (*Heterodera rostochiensis* Woll.) Besides nematode resistance, this species is endowed with a number of other valuable characters, such as high yield and fertility, good crossability with varieties of *S. tuberosum*, resistance to aggressive races of potato wart, field resistance to Phytophthora, resistance to certain virus and other diseases, increased contents of starch, protein and vitamin C in the tubers, etc.

At present the potato nematode has spread widely over almost all countries of Europe and Asia and is found in other parts of the world as well. In our country it was found for the first time in 1948 in the Latvian SSR and Kaliningrad province, in 1949 in the Latvian SSR, in 1953 in the Estonian SSR, in 1957 in Byellorussian SSR and then in many provinces of the RSFSR. Its foci are found in other regions also. This is one of the most dangerous parasites. Chemical control against this parasite is not adequately effective and it is very costly. Therefore, at present much attention is paid to the development of nematode resistant varieties. It is particularly effective to grow nematode resistant varieties as part of the complex of chemical and agronomical control measures against this parasite.

Recently a new species of nematode, *Heterodera pallida* Stone, has emerged. So far this species, as well as the more virulent pathotypes of *Heterodera rostochiensis*, have not been detected in our country. Therefore the major attention is given to the development of varieties resistant to the common (A) pathotype of *Heterodera rostochiensis*.

At present most of the nematode resistant varieties involve *S. andigenum*: Antinema, Amex, Cobra, Tondra (FRG), Sagitta, Specula (GDR), Amelio, Amaryl, Intenso, Marijke, Multa, Prevalent, Provita, Prominent, Rector, Saturna, Element, Echud (The Netherlands), Peconic, Wauseon (USA) and many others. In our country the promising hybrids Meta, Yantar', Ariadna and others have been developed by a number of research institutions.

So far only a few specimens from the huge diversity of forms of *S. andigenum* have been used in breeding for nematode resistance. Therefore, the most urgent task is to identify a greater number of promising nematode resistant specimens from this species and include them in the breeding program. In this context, in order to accelerate and increase the efficiency of breeding work it is necessary to use, first of all, specimens with the greatest number of other agro-economically useful characters together with nematode resistance and to ensure the transmission of these characters to the hybrid progeny.

In the collection of the VIR (All-Union Research Institute of Plant Industry) up until now more than 80 specimens of *S. andigenum* have been isolated and in the progeny of these specimens nematode resistant saplings have been identified. Every year the material is screened for resistance to this parasite at the Byellorussian Research Institute of Plant Protection.

As far as yield is concerned, most specimens of *S. andigenum*, under long-day conditions as in Leningrad province, prove inferior to the varieties of *S. tuberosum* that yield only 100–300 g per plant, but specimens are also found which yield not less than 100 g of small tubers per plant. On the other hand, there are specimens that produce more than 500–1000 g of large tubers per plant. In its native place in South America this species yields up to 200 qtl/ha or more.

In most specimens the starch content of tubers was 13–15%, but there were specimens with 20–27% starch in their tubers. The content of crude protein was mostly 1.7–1.9%, but sometimes it was more than 2.5%. Specimens are found with increased (15–17 mg/100 g) and high (above 18 mg/100 g) contents of vitamin C in the tubers.

While developing varieties suitable for mechanized harvesting, no less importance is given to the so-called "strength" of the tubers estimated with the help of a special apparatus and expressed in terms of "coefficient of strength". Specimens with a coefficient of strength of 60 or more are most suitable.

Many specimens identified for resistance to potato nematode are distinguished also for resistance to the very dangerous parasite, the stem nematode, which makes them even more valuable.

Most of the nematode resistant specimens are of Argentinian, Bolivian and Peruvian origin. Below we give a brief description of some of them in regard to their agro-economically valuable characters.

Specimens from Argentina

Simultaneously resistant to potato and stem nematodes: k-2953, during a period of six years of tubers production at the Pushkin Laboratories, did not show any symptoms of virus diseases. It has field resistance to Phytophthora, produces a good yield, with large tubers, has a satisfactory taste, a starch content of 14–18% and a crude protein content of 1.6–2.6%. The coefficient of strength of the tubers is 73. k-3105 is susceptible to virus diseases but has field resistance to Phytophthora, produces an average yield, has a starch content of 16–20%, and a crude protein content of 1.5–1.9%; k-4622 is resistant to potato wart, is an average yielder, with a satisfactory taste, and a starch content of 17–19%. The content of crude protein is 1.7–1.9%.

The following specimens give a good yield of large tubers with a satisfactory taste: k-3891—resistant to potato wart, starch content 17–19.5%, coefficient of strength of the tubers 40.7; and k-4110—starch content 16%, crude protein 1.9%, coefficient of strength 37.6.

Moreover, nematode resistant saplings segregated in the progeny, giving an average or low yield, are also of interest in potato breeding. Specimens with an average yield are: k-4001—taste satisfactory, starch content 10–14%, content of crude protein 1.5–2.1%; k-4006—starch content 17–19.5%, crude protein 1.6%; k-4496—starch content 14–18%, crude protein 1.7–2.4%; k-4623—taste below average, starch content 15–17%, crude protein 1.8–2.2%, coefficient of strength 27.2. Low yielders with small tubers are: k-3895—resistant to the virus of leaf curling, starch content 15–21%, crude protein 1.8–3.6%; k-3168—starch content 17–19%, crude protein 1.7%; k-3272—starch content 16–20%, crude protein 1.7–2.5%; k-3959—taste satisfactory, starch content 13–17%, crude protein 1.9–2.6%.

Specimens from Bolivia

Resistant to potato and stem nematodes, good yielders with large tubers: k-4611—resistant to virus X, field resistance to Phytophthora, resistant to potato wart, taste satisfactory, starch content 14–23%, crude protein 1.8–2.4%, vitamin C 16.4 mg/100 g, coefficient of strength 47.5; k-4615—over a period of six years of tuber production at the Pushkin Laboratories did not show symptoms of virus diseases, taste good, starch content 17–23%, crude protein 1.7–1.9%, vitamin C 9.9 mg/100 g, coefficient of strength 21–47; k-6970—starch content 17.8%, coefficient of strength 28–35.3, in the progeny six out of 15 saplings were resistant to nematodes and one was slightly susceptible.

Specimens with an average yield: k-4614—resistant to potato wart, taste satisfactory, starch content 12–16%, crude protein 1.7–2.2%, vitamin C 16.4 mg/100 g, coefficient of strength 28.6–31.1; k-4629—taste good, starch content 15–17%, crude protein 1.8–2.5%; k-4631—resistant to virus X, taste satisfactory, starch content 16–18, crude protein 1.8–3.6%; k-4635—susceptible to virus diseases, taste satisfactory, starch content 14–16%, crude protein 2.5–3.0%, coefficient of strength 34.2; k-5591—taste good, starch content 17–20.5%, crude protein 1.4%, in the progeny eight out of 14 saplings were resistant to nematodes and one was slightly susceptible.

Apart from these specimens k-3115 is of practical interest as it is a good yielder, with large tubers and field resistance to Phytophthora, starch content 17–23%, crude protein 2.0–2.9%, vitamin C 18.97 mg/100 g, coefficient of strength 25.7–35.2.

Bolivian specimens have an average or low yield, small tubers and nematode resistant saplings segregate in the progeny: k-5576—taste good, tubers of average size, starch content 16–24%, crude protein 1.3–1.7%, coefficient of strength 90.7; k-4711—taste satisfactory, starch content 17.6%, crude protein 1.3–2.0%, in the progeny three out of 15 saplings were resistant to nematodes and one was slightly susceptible; k-5550—tubers small, starch content 13–20%, crude protein 1.7–2.3%; k-5568—taste satisfactory, starch content 17–19%, crude protein 1.6–1.8%, in the progeny three out of 15 saplings were resistant to nematodes and three were slightly susceptible; k-5573—starch content 17.5–18%, crude protein 1.8–2.0%, in the progeny two out of 15 saplings were resistant to nematodes and three were slightly susceptible; k-5579—taste satisfactory, starch content 14–17.5%, crude protein 1.5–1.7%, out of 15 saplings four were resistant to nematodes and four were slightly susceptible.

Specimens from Peru

Resistant to potato and stem nematodes, good yielders, with big tubers and satisfactory taste: k-1776—resistant to virus X, starch content 12–15.9%, crude protein 1.8–2.5%, coefficient of strength 44.3; k-4619—during the six years of tubers production at the Pushkin Laboratories did not show symptoms of virus diseases, starch content 15–19.7%, crude protein 1.5–1.9%, coefficient of strength 74.4–98.1; k-1714—resistant to virus X, tubers of average size, starch content 16–19.6%, crude protein 1.7–2.4%, coefficient of strength 19.3–48.2; k-3564—good yielder, starch content 16%.

Low yielding specimens with small tubers, nematode resistant saplings segregating in the progeny: k-3083—taste satisfactory, starch content 16–18%, crude protein 1.9%; k-4520—starch content 16.7%; k-7217—tubers of average size, taste good, starch content 16–18%, coefficient of strength 34.1;

k-7953—starch content 17–19.6%, in the progeny three out of 15 saplings were resistant and three were slightly susceptible; k-7956—starch content 17–18%; k-7979—starch content 14–15%, in the progeny four of 13 saplings were resistant and four were slightly susceptible to nematodes; k-7997—taste good, starch content 15–17%, crude protein 1.3–1.5%; k-8108—starch content 10–11%; k-8196—starch content 10%, six out of 15 saplings were resistant to nematodes and two were slightly susceptible.

Specimens from Colombia

Good yielder with big tubers, good taste, starch content 18–20%; k-1690—over a period of a few years of tuber production at the Pushkin Laboratories did not show symptoms of virus diseases, resistant to potato wart, crude protein content 1.9–2.4%, coefficient of strength 60.3–83.8; k-3588—crude protein content 1.7–2.1%, vitamin C 16.12 mg/100 g, coefficient of strength 30–37.8.

Low yielding specimens with small tubers: k-8223—starch content 15–18%, crude protein 1.7–1.9%; k-8269—starch content 16–20%, crude protein 1.6%, vitamin C 16.7 mg/100 g, three out of 15 saplings resistant to nematodes and four slightly susceptible; k-5622—starch content 15%, crude protein 1.4%; k-5806—tubers of average size, taste satisfactory, starch content 17.3-21.6%, crude protein 2.2%; k-5859—taste satisfactory, starch content 18–19%, crude protein—1.7–2.0%.

Among the above-mentioned Colombian specimens none was found to be resistant to stem nematodes as well.

Specimens from Ecuador

Resistant to potato and stem nematodes: k-5583—taste good, starch content 16–23%, crude protein 1.6–2.4%, vitamin C 13.76 mg/100 g, coefficient of strength 16–22.

Thus for use in breeding nematode resistant varieties of the potato there is a good reserve of new initial forms within the limits of *S. andigenum* which had not previously been used by breeders. Their involvement in hybridization opens new possibilities for breeding nematode resistant varieties of the potato.